How to cut $2 Trillion

MW01701158

Print ISBN 9781630692940
Digital ISBN 9781630692933

Published by Brownstone Institute 2025

Creative Commons Attribution 4.0 International

How to cut $2 Trillion

A Blueprint from Ronald Reagan's Budget Cutter to Musk, Ramaswamy, and the DOGE Team

DAVID A. STOCKMAN

BROWNSTONE
INSTITUTE

CONTENTS

CHAPTER 1

America's Fiscal Doomsday Machine Must Be Stopped

The DOGE $2 trillion budget savings goal is crucial to the very future of constitutional democracy and capitalist prosperity in America. In fact, the soaring public debt is now so out of control that the Federal budget threatens to become a self-fueling financial doomsday machine.

Recall this sequence. When Ronald Reagan was elected in 1980 on a call to bring the nation's inflationary budget under control, the public debt was $930 billion and about 30% of GDP.

By the time Donald Trump was elected the first time it had erupted to $20 trillion, which has now become $36 trillion and 125% of GDP. Moreover, by the end of this decade the Federal fiscal equation will be going supercritical without sweeping budget reductions at the level of the DOGE target. Thus, by FY 2034 the annual baseline deficit according to CBO will total $2.9 trillion and 7% of GDP.

Yet even these enormous figures are based on a Rosy Scenario fairy tale. Namely, that Congress will never again adopt another spending increase or tax cut, including the impending $5 trillion extension of the expiring 2017 Trump tax cuts. It also conveniently assumes there will be no recessions, no inflation recurrence, no interest rate flare-ups nor any other economic crises for the remainder of this decade and forever thereafter.

Furthermore, it presumes that these surging red ink totals and soaring debt service expenses would be copacetic in the bond pits just the same. That is, CBO inexplicably projects that 7% of GDP deficits and annual interest expense of $1.7 trillion or 4.1% of GDP by 2034 would be compatible with a weighted average yield on nearly $60 trillion of public debt of just **3.4%**.

Yes, and if dogs could whistle the world would be a chorus! Give the average yield just another 250 basis points, however, and now you have $3.1 trillion of annual debt service expense and a $4 trillion annual deficit by 2034. In short, there is a doom-loop building inside the Federal fiscal equation and nothing short of the DOGE target of $2 trillion of annual budget savings by the end of this decade can reverse its explosive materialization in the years beyond.

If sweeping budget retrenchment does not occur soon, in fact, soaring interest expense will ignite a veritable fiscal wildfire. On paper, the public debt would power upward unabated to ***$150 trillion*** or 166% of GDP by mid-century (2054) under CBO's current Rosy Scenario projections. Of course, long before the debt actually hits this staggering figure, the whole system would implode. Every remnant of America as we now know it would go down the tubes.

So we need to be clear that the DOGE team of Musk and

Ramaswamy must focus on savings of $2 trillion *per year* commencing relatively soon. That's because the nation's fiscal doomsday machine will be accumulating interest expense so fast as to make $2 trillion of savings spread over a longer period–such as a decade–little more than a rounding error. To wit, Federal interest expense has already passed the $1 trillion per year mark, will exceed $2 trillion per year in the early 2030s and would top *$7.5 trillion per year* at minimum by our calculations by mid-century.

Stated differently, if something drastic is not done now – like a $2 trillion ***annual*** budget savings by the end of Donald Trump's second term – America will be paying more interest on the public debt within 25 years than the entirety of today's Federal budget. That's right: Debt service will exceed current outlays for Social Security, defense, Medicare, education, highways, the national parks, Head Start, interest, and the Washington Monument, too.

Obviously, the sprawling Federal government and its prodigious expanse of spending and debt literally defies easy comprehension and graspable solutions. After all, the current annual budget of $7 trillion amounts to Federal spending of nearly $20 billion per day and $830 million per hour. And when you talk about the 10-year budget outlook, comprehension literally fades away completely: The current CBO spending baseline for 2025-2034 amounts to ***$85 trillion*** or just shy of the *annual GDP of the entirety of planet Earth this year.*

So based on experience we suggest that the DOGE team needs to build its $2 trillion case around a target year and several big buckets of savings by broad type. The latter can then be used to fashion a detailed but comprehensible blueprint for arraying and conveying the desperately needed housecleaning of the Federal budget that the DOGE has been tasked with accomplishing.

In that context, FY 2029 makes the most sense as a target year since it would represent the 4th and outgoing Trump budget; and also one which would give sufficient time for phasing in some of the sweeping cuts that will be needed, but not so far in the distant future as to be largely irrelevant to the here and now of fiscal governance during Donald Trump's second term.

We'd also suggest three big buckets of savings, which we would short-hand as follows:

- ***Slash the Fat....***by eliminating unnecessary and wasteful agencies and bureaucrats wholesale.
- ***Downsize the Muscle....***by curtailing national security capacities and functions that have grown up during the Forever Wars but are not needed for an America First foreign policy.
- ***Cut the Bone....***by reducing low-priority entitlements and subsidies that the nation cannot afford, and which a reasonable view of societal equity does not require.

Needless to say, when it comes to the vast wasteland of the Federal budget there are innumerable ways to skin the cat. But based on our own experience of more than a half-century of familiarity with the Federal budget as both a participant and an informed observer, we judge the following mix to be the most plausible and balanced way to get to the $2 trillion of annual savings by FY 2029.

To be sure, even this relatively judicious mix is sure to ignite firestorms on the banks of the Potomac like never before, but it can be strongly justified and defended for the reasons we will lay out in detail below.

Annual DOGE Savings Targets By Component:

- Slash the Fat: $400 billion or 20%.
- Downsize the Muscle: $500 billion or 25%.
- Cut the Bone: $1.1 trillion or 55%.

Suffice it here to say that the first bucket alone would leave them screaming to high heaven in the swamplands of DC. But even that $400 billion savings could be accomplished only by eliminating 16 agencies entirely, slashing another nine departments by 50%, cutting the balance of the nondefense payroll by 34%, terminating $40 billion per year worth of wasteful farmer subsidies, cancelling entirely $60 billion per year of energy boondoggles including all EV credits, and eliminating $150 billion per year of all other forms of corporate welfare and subsidies embedded in the budget and tax code.

We will amplify the details of this $400 billion of inherent Federal budget fat and waste in the chapters below. But suffice it here to say that attacking the usual shock effect lists of outrageous studies, stupid foreign aid projects, or even payments to dead people, as is often used to illustrate wasteful spending, will get you barely a fractional decimal point of the savings target, as desirable as eliminating this nonsense might be in its own right.

For instance, a recent "outrageous spending" list showed $4 million was wasted on "Dr. Fauci's Transgender Monkey Study" and $6 million on a "USAID Fund to Boost Egyptian Tourism," among countless more absurdities. Still, eliminating these two items would contribute only ***0.0005%*** to the $2 trillion savings target.

Even some of the larger ideas of this sort, such as timelier elimination of dead people from the Social Security rolls, would not get you very far, either. To be sure, 1.1 million Social Security recipients pass on to their rewards each year, while departing beneficiaries would be receiving an average benefit currently of $1,907 per month. So one extra month of dead people on the rolls costs the not inconsiderable sum of ***$2.1 billion.***

At the present time, however, not much excess dwell time actually happens. The rolls are purged every month based on newly filed death certificates, and this encompasses the termination of payments to anyone who died during the course of the month, including the last day. So the average duration on the rolls of Social Security decedents is 15 days, which computes to $1.050 billion of payments.

Of course, if the Musk and Ramaswamy team could come up with some more super-duper software to monitor, report, calculate final month benefits and then terminate decedents in real time, it might reduce dwell time by two-thirds. In turn, this means that getting dead people off Social Security 10 days faster would generate a savings of $700 million per year or about *0.04%* of the $2 trillion target. That is to say, there is undoubtedly room for efficiency improvements and elimination of outright waste and stupidity everywhere in the Federal budget, but it unfortunately adds up to rounding errors.

Stated differently, if it doesn't "scream and bleed" politically it won't likely make a dent in achieving the $2 trillion goal. There is just plain nothing antiseptic about slashing the Federal budget.

For instance, even a thundering ***50% cut in the current nondefense Federal headcounts*** of 1.343 million would save just $100 billion annually by the target year of 2029. And that's a comprehensive figure based on the current average cost per Federal employee of $100,000 in pay per year plus $44,000 in average benefits and fringes--escalated for inflation to $160,000 per bureaucrat by FY 2029.

Accordingly, to reach $2 trillion of annual savings will require a deep dive into the three buckets listed above. In the next five chapters we will lay out the most plausible and judicious route to the $400 billion of "Slash the Fat" savings, followed by the details and an America First rationale for cutting $500 billion per year of unneeded muscle from the national security budget in Chapter 7. Chapter 8 will then delve into $1.1 trillion per year of cuts from the bone of entitlement and domestic welfare that would be needed to reach the $2 trillion DOGE savings target.

But one thing should be clear from the outset. Lists of outrageous anecdotal items provide color about the stupidity and waste that is rampant in the Federal government. But they have nothing whatsoever to do with the fact-based analysis and philosophical U-turns that will actually be required to complete the DOGE mission successfully.

CHAPTER 2

Slash The Fat: 16 Agencies To Terminate

Under our three savings bucket scheme, "Slashing the Fat" from the Federal payroll and bureaucracy would account for just ***$400 billion*** or ***20%*** of DOGE's $2 trillion per year savings target. Needless to say, however, even that small portion would be far easier said than done.

That's because, unlike the case of typical US businesses, where payroll costs can range from 15% to 40% of total costs, such expenses comprise only a tiny fraction of total Federal spending. Setting aside DOD payrolls for the "Downsize the Muscle" bucket, we estimate fully-loaded nondefense employee compensation costs at ***$215 billion*** in the target year of FY 2029. That's just *3.1%* of the $7 trillion of nondefense outlays projected under current policy by CBO for what would be the final Trump budget.

So there is a lot of wood to chop in other areas of nondefense spending, but we start with the assumption that ***$85 billion*** or 40% of nondefense payroll costs would be a fair component of a broader plan to generate the $400 billion of "Slash the Fat" savings. At the projected FY 2029 cost of $160,000 per Federal employee for payroll, benefits, and fringes, this would require termination of ***535,000*** positions from the current total of 1,343,000 nondefense employees.

On its face, this headcount reduction target is eminently plausible given that the Washington Swamp is a vast cesspool of padded payrolls, useless projects, endemic inefficiency, and misbegotten government enterprises. But what is especially telling is that our ***40%*** payroll cut would amount to just half of the ***80%*** staff reduction that Elon Musk achieved at the old Twitter. And he did so in the context of a labor-intensive business without missing a beat in terms of operations and customer accommodation at the new "X."

So we begin the payroll savings analysis by bringing the hammer down terminally on the 16 worst and most unneeded Federal agencies, including the FBI, OSHA, the FTC, and the Department of Education. Eliminating these 16 bureaucracies entirely would reduce Federal employment by ***71,000*** jobs and save ***$11.1 billion*** per year of direct compensation costs. That's nothing to sneeze at, of course, but to place it in budgetary context it does represent ***only 13 hours' worth*** of the $8.0 trillion per year of total baseline Federal spending for the target budget year of FY 2029.

We also show that cutting 50% of the staff levels at another 9 dubious departments–including the EPA, NASA, and GSA–would shrink the Federal payroll by an additional ***93,000.*** That would save a further ***$15 billion*** annually in compensation costs.

Still, we would need an additional ***$59 billion*** in nondefense savings to achieve the $85 billion target for direct compensation reductions. Accordingly, upwards of ***371,000*** positions would need to be eliminated from the balance of the nondefense agencies or about 34% of the 1,084,000 current jobs at everything from the Agriculture Department to the Social Security Administration and Veterans health care system.

In addition, we estimate that $85 billion in compensation cost savings would generate an additional $45 billion of indirect savings in the related costs for agency overhead, occupancy, supplies, and outside contractor services.

In summary, therefore, we'd propose that about one-third of the "Slash the Fat" savings target of $400 billion be obtained from the following areas *inside the four walls* of nondefense government. Chapter 6 will also outline $270 billion of savings from outside the walls of nondefense government in the form of cuts in corporate welfare, farmer subsidies, the Green New Deal, and other wasteful private sector subventions.

Summary of Savings From Headcount and Nondefense Agency Waste Reductions (FY 2029):

- 100% Elimination of Staffing at 16 Unnecessary Federal Agencies: $11 billion.
- 50% Staffing Cut at 9 Dubious Federal Agencies: $15 billion.
- 34% Staff Reduction at All Other Nondefense Departments: $59 billion.
- Indirect Overhead savings from nondefense staff reductions and agency eliminations: $45 billion.
- Total Nondefense Staff and Overhead Savings: ***$130 billion.***

We begin with a summary of the 16 agencies to be shut down, along with the number of staff positions to be eliminated and the resulting direct employee compensation savings. These agencies are slated for complete elimination because in the context of a roaring fiscal crisis, they are either utterly unnecessary or inappropriate functions of government or comprise activities that are already being handled by other Federal agencies, state and local governments, or the private sector.

Self-evidently, these 16 agency closures would result in only a small down payment against the $2 trillion per year savings goal. Yet it is crucial to start here because each of these agencies represent cases of egregious regulatory excess or Washington-based enterprises that are not remotely the business of the central government in any season, but most especially not during a time when the Federal government is careening toward the fiscal shoals.

Stated differently, the list below comprises a kind of Litmus Test of fiscal resolve. If these Federal bureaucrats and agencies can't be eliminated, the prospect for reining in America's unfolding fiscal calamity is dim indeed.

16 Agencies To Be Eliminated–Staff Cuts and Payroll Savings:

- National Endowment for the Arts: 100 staff and $16 million savings.
- National Endowment for the Humanities: 100 staff and $16 million savings.
- Legal Services Corporation: 800 staff and $128 million savings.
- National Highway Traffic Safety Administration (NHTSA): 600 staff and $96 million savings.

- Federal Trade Commission (FTC): 1,125 staff and $180 million savings.
- Corporation for Public Broadcasting: 100 staff and $16 million savings.
- OSHA: 2,200 staff and $352 million savings.
- Consumer Products Safety Commission: 600 staff and $96 million savings.
- Agency for Global Media: 1,125 staff and $180 million of savings.
- National Endowment for Democracy (NED): 162 staff and $26 million savings.
- Education Department: 4,245 staff and $680 million savings.
- Consumer Financial Protection Bureau: 1,500 staff and $240 million savings.
- Agency for International Development (AID): 10,000 staff and $1.6 billion savings.
- FBI: 34,000 staff and $5.4 billion savings.
- BATF: 5,300 staff and $848 million savings.
- DEA: 9,315 staff and $1.49 billion savings.
- Total 16 Agencies To Be Eliminated: ***71,000*** staff and ***$11.3*** billion savings.

As it happens, many of the above listed agencies were on the original Reagan zero-out list of 1981. Yet they are still alive and prosperous because the Swamp is relentless in the defense of its own, and especially because on the margin even most of the GOP movers and shakers on the Congressional spending committees have been Washington lifers, RINOs, and political weaklings afraid to resist the politically correct dictates of the Washington establishment and their megaphones in the MSM.

National Endowments for the Arts and Humanities

For instance, they are still spending about $420 million per year on the National Endowments for the Arts and the National Endowment for the Humanities. Back in 1981, when the public debt was still under ***$1 trillion*** and 31% of GDP, we argued that the cultural institutions supported by the endowments should be funded by private philanthropy and public admission tickets to museums, operas etc., not hard-pressed bus drivers in Milwaukee struggling to feed, clothe, and shelter their families; and most certainly not via borrowing from future taxpayers through endless deficit finance.

At the time, the net worth of the top 1% of households was about ***$3 trillion***, indicating ample capacity among wealthy patrons to support America's important cultural institutions and endeavors, along with the voluntary support of millions of other less prosperous but culturally engaged citizens.

Well, here we are 44 years later with a public debt of ***$36 trillion*** and heading skyward, while the net worth of the wealthiest 1% of US households has risen by ***16X*** to ***$47 trillion.*** And that staggering wealth pile stands alongside an additional current net worth of $10 trillion for the next 9% of wealthiest households. Yet and yet: The clueless politicians on the Potomac are still borrowing money to fund cultural institutions when the top 10% of US households alone have *$56 trillion* of net worth available for support of the arts and humanities.

In this instance, we'd suggest that Elon Musk set the example by pledging $2 billion over the next five years to enable cultural institutions and artists time to locate alternative sources of funding, thereby permitting the national endowments to be zeroed out immediately. This would at least get the agency-elimination ball rolling with a bang!

To be sure, shutting down the two endowments would result in a reduction of only 200 Federal jobs and a compensation cost savings of just $32 million per year, but as we will detail below, it would also generate additional savings from grants and overhead of $390 million.

In any event, this is surely the place to start. After all, if the Trumpified Washington can't even eliminate these two agencies, why then, truly, all is lost.

Legal Services Corporation

The same goes for the 800 staffers and $128 million of savings from eliminating the Legal Services Corporation. For crying out loud, this whole operation is a liberal hobby horse dating back to the early days of the War on Poverty in 1965.

If the dubious political litigation it mainly supports via direct staff and another $432 million of grants and contracts has not found non-Federal funding more than a half century later, it doesn't deserve another dime from Uncle Sam. Period.

National Highway Traffic Safety Administration (NHTSA)

In the case of the NHTSA, we have the very worst of the Nanny State. It has not only usurped the role of the private market and legal liability system in determining appropriate engineering standards for auto safety, but for decades has been knee-deep in setting idiotic average fuel economy standards (CAFE) for the *entire fleet* of each automaker.

This causes immense distortions in vehicle offerings, pricing, and production sourcing. That's because to meet fleetwide fuel economy mandates each automaker must average together the lower fuel economy ratings of heavier, higher performance and profitable vehicles the public actually wants to buy with the artificially high fuel economy levels of small, stripped-down, under-powered cars that they must be deeply discounted to move the metal owing to limited marketplace appeal. In the process of compliance, automakers also tend to shift sourcing of the latter small, cheap "compliance" vehicles to Mexico and East Asia in order to relieve the strain on profitability resulting from these largely unprofitable NHTSA-mandated autos.

Accordingly, we would propose to abolish the NHTSA and in one fell swoop get rid of 600 bureaucrats and an overall waste of $1.2 billion per year, including about $500 million of safety grants to the states. With respect to the latter, if the genius socialist legislators in Sacramento and Albany want to steer their own unwashed driving masses to purportedly safer modes of happy motoring, let them do so on their own taxpayers' dime.

Abolition of the NHTSA would also return consumer vehicle choice to the marketplace and likely bring a lot of current foreign-sourced auto production back home. That is to say, most of today's auto companies–both the Big Three and foreign brands–make a decent profit manufacturing full-sized sedans, SUV's, and pickups in the United States. Upon abolition of the CAFE program, therefore, Nanny State-mandated and foreign-sourced econo-boxes would lose their helping hand from Washington, paving the way for more US built vehicles on dealer lots that consumers actually wish to buy.

And, yes, if consumers want six airbags per car as now mandated by the NHTSA (standard sedans are required to have two frontal airbags, two side airbags and two curtain airbags to protect occupants in the event of a side-impact crash), manufacturers will offer dealer-installed options at the appropriate (steepish) markup to base sticker prices. Indeed, the idea that consumers need a Federal Auto Nanny in order to choose a "safe" vehicle goes back to Ralph Nader's original grab for regulatory power back in the 1970s and 1980s, which we fought in Washington when at least some Republicans still understood the statist scam of alleged "market imperfections."

Federal Trade Commission

America imports $3.1 trillion of goods every year, which is testimony in itself that planet earth is crawling with potential competitors, fair and unfair. This actual and potential competition militates against the ability of any domestic manufacturer to monopolize anything.

In fact, students of sound market economics have understood since at least the 1960s the populist idea that private capitalism is an incubator of monopoly is just plain nonsense. With extremely rare exceptions, monopolies and rigged oligopolies only arise when they are *enabled by the state* via regulatory favoritism and capture, subsidies, and/or protectionist restraints of both domestic and international trade.

So what Washington needs is not anti-monopoly policemen, but the elimination of crony-capitalist policies that bestow unfair and coercive competitive advantage on politically privileged competitors. Most certainly, therefore, two anti-monopoly bureaucracies are way beyond the ken, meaning that the FTC should be abolished entirely. If need be, any minor residual meddling with business in this area can be handled by a low-cost rump operation in a drastically downsized antitrust division of the DOJ.

Again, savings of $180 million per year of FTC compensation expense is more than warranted, even as it would free American business from Nanny State meddling that results from 1,125 FTC staffers scurrying around in search of imaginary problems to justify their salaries. And, as we will amplify below, there would be a bonus savings here of $250 million, representing the non-payroll waste incurred by the FTC.

Corporation for Public Broadcasting

Even back in the world of 1981, there was no case for public funding of radio and TV, but by the year 2024 it has become a screaming instance of "Oh, puleeeze!"

The powerful presence of "X" (nee Twitter) is testimony itself that the dominant hometown newspaper and three broadcast networks no longer have even a remote monopoly on the news. That was the ostensible reason for the government-funded NPR back in the day, which, predictably, was bypassed by the flowering of tens of thousands of technologically and market-based alternative media and news/information/entertainment venues. And then, even as NPR became redundant and utterly unnecessary, it morphed into a state propaganda agency to boot.

Accordingly, the CPB's 100 staffers should be told to send their resumes to the blooming, buzzing world of alternative media on Day 1, even as the expense of $16 million for compensation and $520 million for affiliate grants and contracts is eliminated. Cold turkey would be the obvious way to serve up the savings in this case.

OSHA (Occupational Health and Safety Administration)

As it happens, there are approximately 90,000 units of state, county, city, village, and township government in the United States–the overwhelming share of which are involved in the business of grassroots public health and safety administration and enforcement in some form. So, if these manifold units of government can't look after safety in the workplace–from farms to warehouses and factories–what's even the point of the Founders' genius? Namely, their acute understanding that healthy democracy requires a decentralized federalist form of the state, not a unitary power in a capital city distant from the daily life of the people and the marketplaces and communities in which they operate.

Beyond that, there is no absolute science of workplace safety. Always and everywhere, it involves a trade-off between levels of protection and costs, and also choices among an infinite array of engineering versus behavior approaches to safety–all of which have their pros and cons. That's why a federalist approach is tailor-made for the very function and jurisdiction of OSHA.

That is to say, Justice Brandeis had the answer more than a century ago when he argued that the states were the proper laboratories of democracy and that many of the functions Washington has since usurped might be better experimented with and executed at the state and local level.

In the case of cowboy safety, for example, the California-style approach illustrated below might be appropriate for a state that lost its cowboys long ago, anyway. But Texas, which still has some, might well prefer a more practical and less burdensome approach.

In any event, ***2,200 bureaucrats and inspectors*** on the OSHA payroll are absolutely unnecessary to insure safe workplaces in America. Not only would the elimination of OSHA save $350 million of staffing costs and $1.3 billion of annual Federal expense overall, but it would also relieve businesses and workplaces in America of literally billions of compliance costs and millions of hours of paperwork that represent the inherent overkill of a centralized bureaucracy that has become a captive of its own labor union constituencies.

Besides, we'd bet that Florida, the Carolinas, and Texas would be more than happy to accommodate the relocation of businesses that might be chased away by a mini-OSHA in Albany, Sacramento, or Springfield. That is, competition among the states for investment, jobs, and a favorable business climate is likely to be a far more powerful brake on regulatory agency excesses than the so-called Congressional oversight committees ever have been, or even the courts–neither of which have real skin in the game.

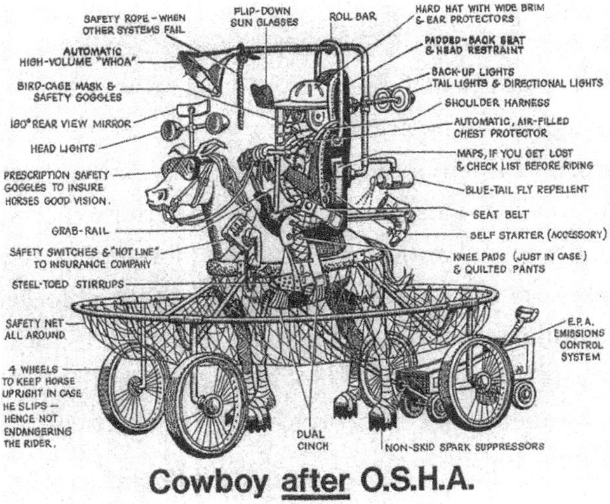

Cowboy after O.S.H.A.

Consumer Products Safety Commission (CPSC)

Even more than OSHA, the Consumer Products Safety Commission is a case of the Nanny State run wild. When you look at the main product categories of its regulatory focus listed below, you have to wonder how in the world American consumers even dared go into a furniture mart, hardware store, or children's toy emporium without risking the life and limb of themselves and their families before CPSC's enactment in 1972; and also what the other 90,000 units of state and local government were doing with respect to the very prosaic matter of household product safety–to say nothing of parents and grandparents.

As to the latter, we have fond memories of a 12-foot-high swing our grandfather rigged up from the high limb of a big Maple tree in our backyard. He was undoubtedly not CPSC-compliant in his swing making, but he damn well knew what was safe for kids and therefore secured the ropes and seat far more safely than what happened when we kids used his swing to "bail out" in imitation of fighter pilots exiting a burning plane.

Then again, the free market has a very powerful incentive for vendors to make and sell safe products: Namely, the protection of their brand franchises and avoidance of devastating legal liability settlements for defective products, which settlements in today's world can badly impair or even bankrupt a careless or crooked business. After all, the torts bar was a powerful health and safety line of defense for consumers long before the Nanny State even arose.

In any event, as in the case of workplace safety there is no "science" whatsoever with respect to "safe" baby cribs, adult mattresses, power drills, deodorants, or ATVs. It's all a matter of trade-offs between cost and functionality, on the one hand, versus product safety, on the other. It also involves complex issues of engineering versus behavior-based risk mitigation, and ultimately turns on consumer preferences and risk propensities.

For instance, the "sport" of skydiving is both hazardous and fully legal, but an inherently safer ATV must have CPSC-compliant roll bars, seat belts, helmets, safety instruction manuals, and speed governors on models designed for younger drivers.

Indeed, if any of the products listed below actually require state-imposed regulation reaching beyond the inherent protections of liability law, there is still absolutely no reason to supersede the ample net of traditional regulation by state and local governments, trade associations, and insurers of product liability risk that existed before 1972.

Yet this very observation tells you all you need to know about Nanny State regulation. To wit, the CPSC has thrived politically over the decades since 1972 because crony capitalists have learned to love regulation from the Washington Swamp. Very simply, it avoids the inconvenience and costs of meeting different regulatory standards in California versus Utah and Indiana; makes for one-stop lobbying on K-Street; and creates barriers to entry for upstart competitors.

Then again, it is not the legitimate business of the Federal government to protect American businesses from the foolishness of regulatory zealots in the Socialist Republic of California or the New York Soviet in Albany. Once again, in fact, the only practical way to minimize wasteful and costly regulatory interference with the production and sale of the myriads of everyday consumer products listed below is via *energetic competition between the states*.

We are quite confident that the likes of Utah, Kansas, Tennessee, and Florida would find the right balance with respect to the safety of cribs, toasters, bicycles, and camping gear long before consumers were forced to march on Sacramento seeking regulatory relief from California-mandated high-cost and low-performance versions of these same products.

Functions of the CPSC:

- Ensuring the safety of toys, cribs, strollers, and other children's items.
- Regulating items like furniture, mattresses, and household appliances to prevent injuries from fires, falls, and electrical hazards.
- Ensuring the safety of sports equipment, bicycles, and playground equipment like slides and swings.
- Ensuring that consumer electronics, including small appliances and power tools, meet safety standards to prevent electrical shocks and fires.
- Regulating household chemicals, cosmetics, and personal care products to reduce risks of poisoning, burns, and other injuries.
- Overseeing the safety of items like ATVs, boats, and camping equipment.

Agency For Global Media

The Congressional foreign affairs and national security committees thrive upon globetrotting junkets and strutting around foreign lands as visiting plenipotentiaries of the

American Empire. So they have found it inconvenient to acknowledge that the Cold War ended 34 years ago and that many of the institutions erected to fight it are now utterly obsolete, if they were ever necessary in the first place.

The prime example of this would be the string of US government propaganda agencies including the Voice of America, Radio Free Europe/Radio Liberty, Radio Free Asia, and the Middle East Broadcasting Network. These were all designed to counter exaggerated claims that the commies were on the march toward global domination and that the backward peoples of other nations needed to be educated as to this danger by enlightened patriots bivouacked on the banks of the Potomac.

Of course, the commies are now long gone. Well, unless you think the Red Capitalists of Beijing are really mobilizing a vast armada of 100 aircraft carriers and troop ships to disembark on the shores of California, thereby destroying their own global export trade, Ponzi-economy, and basis for continued rule in the process.

To the contrary, of course, the picture is more nearly the opposite: The calamity of Vietnam, two pointless but bloody and destructive wars in Iraq, the veritable carnage visited upon Syria, Libya, and Yemen by American arms and the missiles and bombs stamped "Made in the USA" now raining down from the skies in Gaza, Lebanon, and Ukraine surely leave serious doubt about whether any of these obsolete propaganda agencies are convincing the unwashed masses anywhere on the planet to bend a knee toward Washington.

In any event, an America First national security policy focused on an invincible nuclear deterrent and safeguarding of the North American shorelines and airspace from conventional attack, as we will outline under the "Downsize the Muscle" basket in Chapter 7, doesn't need to waste a single dime on the 1,125 bureaucrats employed by the parent agency of these Cold War relics. And that's especially the case in an internet-saturated world where even the muscular dictators of Beijing and their Great Firewall of China can't effectively suppress unauthorized communications emanating from outside the Middle Kingdom.

Abolishing the Agency for Global Media would thus save $180 million per year in direct compensation costs and an additional $770 million wasted on contractors, facilities, communications equipment, and rentals, etc. In a world with today's communications technology, in fact, America is inherently, for good or ill, the proverbial light on the hill.

That's because anything and everything which transpires here–from sea to shining sea–is transparent to the entire planet. The world now sees on the internet all that we do in real time and makes its own judgment. Washington doesn't need to waste dollars it doesn't have on the salaries of journalism school graduates peddling Warfare State propaganda in the process of building up their own resumes for more lucrative opportunities in the MSM.

National Endowment for Democracy (NED)

The 162 staffers and $315 million annual budget of NED are not only a waste, but also a purely destructive project of the Washington neocons and warhawks. We fought it tooth-and-nail when it was concocted in 1983 by neocons in the Reagan White House, arguing it would become a sinecure for Washington national security lifers who didn't make the big grades at the CIA, State, and DOD.

We were right about that in spades. A former president of the Young People's Socialist League (YPSL), Carl Gershman, became its first executive director in 1984 and was still there in 2021, when they finally gave him his gold watch. But like all former Trotskyites who turned neocon under the tutelage of the detestable Irving Kristol and his equally reprehensible son, Bill Kristol, Gershman spent the 37 years of his tenure carrying out the CIA's regime-change function that was seconded to NED in the 1983 legislation.

Among all of the "color revolution" follies promoted by NED during these years, the most insidious was its role in organizing, funding, and enabling the Maidan uprising in Kiev during February 2014. That pointless exercise in regime change paved the way for the Washington-fostered putsch which installed neo-Nazi-sympathizers and militant Ukrainian nationalists in power in Kiev.

In turn, Washington's illegal deposing of the legitimately elected Russian-speaking and Russian-sympathizing president, Viktor Yanukovych, who had won office in 2010 on the back of massive 80%+ margins in the Donbas, Crimea, and Black Sea rim, paved the way for the current civil war carnage, and disastrous proxy war on Russia. After all, the Ukrainian nationalists who were picked, named, and recognized by Washington quickly put joining NATO into Ukraine's constitution, outlawed Russian language, and launched a brutal civil war against breakaway Russian-speaking regions, thereby eventually provoking the Russian invasion of February 2022.

Since then, the US has spent upwards of $150 billion on

a pointless war of human and infrastructure destruction–a veritable demolition derby of senseless military intervention. And one which now threatens to bring Washington's reckless proxy attack on Russia to the brink of nuclear confrontation. ***Yet the Ukraine disaster is the very quintessential work of NED. That alone merits its abolition–no more questions asked.***

But, alas, there is another point. More than half of NED's annual $300 million of taxpayer money is used to sustain the worst kind of Washington insider corruption, log-rolling, and self-justifying promotion of the Warfare State. To wit, half the funds are divided up among institutions controlled by the four big political powers which operate on the banks of the Potomac. That is, the union promoted "American Center for International Labor Solidarity," the business sponsored "Center for International Private Enterprise," the Democrat-controlled "National Democratic Institute for International Affairs," and the GOP-controlled "International Republican Institute." The purpose of these beltway duchies, of course, is to fund cheerleaders for the projects of Empire abroad.

Moreover, for good measure, the balance of the $300 million goes to hundreds of NGOs based abroad. These are essentially the advance guard of Washington's Empire First policies and should not get a dime under a regime of America First.

It can be well and truly said, therefore, that there is no conceivable waste more egregious and more rotten than that embodied in NED. It needs to be shot deader than bin Laden at the very earliest opportunity.

Education Department

Needless to say, the Education Department should never have been potted on the banks of the Potomac because education is meant to be a state, local, and parental function across the length and breadth of the land. Indeed, centralization and national dictation of educational processes, standards, content, and institutional arrangements is the very last thing that should come under the control of the central state.

The current education department came into existence, in fact, only in 1979 as a desperate Carter administration sop to the teachers' unions, which were the backbone of his political coalition. Accordingly, an immediate shutdown of this still infant and unneeded department was a high priority on the Reagan administration's zero-out list.

As it happened, however, squirrely GOP politicians on the Congressional education committees and a Secretary of Education who spent his time in office sabotaging the president's budget prevented the Department from being strangled in the cradle as intended. Instead, the education lobby's victory over the Reagan challenge enabled the new department to thrive without interruption for the next 40 years until reaching a monstrous $350 billion spending level at present.

Still, there is only one way to ensure free expression, diversity of pedagogical approaches, and unfettered experimentation in the education sector. To wit, abolish the Department of Education completely, spin off existing Federal grant activities in block grants to the states, at a reduced percentage of existing funding levels, and slash subsidized student aid by 40%, as we outline in Chapter 8.

Actually, this is not mission impossible. During 2024 the

broad allocation of the Department of Education's funding was as follows:

- Elementary and secondary education grants and support: $52 billion.
- Special Education, Adult, Career, and other Education programs: $18 billion.
- Higher Education Pell Grants, Work Study, and other student direct aid: $30 billion.
- Subsidized Student Loan Outlays: $250 billion.
- Total, Federal Education Programs: $350 billion.

Needless to say, DOGE could dispense with the Department of Education's 4,245 bureaucrats and their $680 million compensation costs in one fell swoop by proposing to eliminate the department entirely. Yet the educational institutions of America would be no worse for the wear if the massive array of programs and activities now administered by the department at huge overhead costs were to be bundled up into block grants and distributed on a no-strings basis to the states in *direct proportion to each state's share of Federal taxes*.

Thus, the first two lines above could be combined in the form of an "Elementary and Secondary Education Block Grant" and funded at 70% of current levels or $49 billion per year, while the dozens of programs under the third line would be packaged into a "Higher Education Block Grant" at an initial level of $18 billion. Since both block grants would represent a pure return of Federal taxes paid by the states' taxpayers, the block grants could be phased out progressively over the ten years after 2029–an ample period of time to permit the states to tax and fund their own education programs or return the money to taxpayers, as they so choose.

Finally, the very idea of "student loans" is flat-out ridiculous because in today's dynamic world it is virtually impossible to solvently underwrite the value of a higher education. And that's true whether in advanced mathematics or basket-weaving, as the case may be. Indeed, as the recent Biden reelection-driven loan forgiveness ploy reminds, "student loans" are essentially nascent welfare grants waiting for opportunistic politicians to cancel the repayments.

This whole area of student financial assistance therefore belongs entirely in the realm of *income transfer payments* and social redistribution. If the latter is to be done at all, it should preferably be based on economic need and structured according to the lights of voters and their legislators in the several states. So if California wants to offer college students a bonanza of subsidies, it should ask its own taxpayers to foot the bill.

In any event, abolishing Federally funded student loans would mainly reduce the current deep implicit Federal subsidies to the affluent upper middle class and the rich who owe most of the $1.74 trillion of student loans outstanding. That's a most worthy goal in its own right, while the favorable FY 2029 and longer-term budget impact will be amplified in Chapter 8.

Consumer Financial Protection Bureau (CFPB)

Never has a more unnecessary and utterly wasteful agency being created in Washington than the CFPB, which was a conscience-relieving sop to Congressional liberals and other Beltway politicians concocted by Senator Chris Dodd and

Congressman Barney Frank to atone for Congress' egregious $700 billion TARP bailout of Wall Street. Yet the so-called Great Financial Crisis had been caused by reckless mortgage and housing market ***speculation*** enabled by the Fed's hideously low interest rates, not sharp practices at retail banking windows.

Even the abuses of so-called liars' loans and other scams in the home mortgage market were a function of easy money and lax oversight by bank supervisory agencies, not because mortgage borrowers got tricked into lying about their income or assets!

So there was exactly zero reason to stand up a new $650 million per year regulatory agency sporting 1,500 more bureaucrats to protect financial services consumers. Well, except to humor Congressional grandees like Dodd and Frank and their GOP co-conspirators on the other side of the aisle.

Indeed, the CFPB's stated mission "to protect consumers in the financial marketplace by ensuring transparency, fairness, and accountability" is outright nonsensical. The fact is, due to regulatory protections and generous government subsidies like FDIC insurance the US economy is vastly *overbanked.*

There are now roughly 5,400 banks and thrifts holding $24 trillion of assets–along with 4,600 credit unions, 240 money market funds, and a proliferating array of online nonbank alternatives that expand by the day. And all of these institutions are hungry for business and aggressively compete for customers. One bank's sharp practices, therefore, is the next bank's sales pitch as to why it is more trustworthy and reliable.

So it's time Washington finally recognizes that the consumer's best and ultimate protection is the competitive free market and that today's financial system is prodigiously endowed with exactly that. Consumers plainly do not need a Financial Nanny on the banks of the Potomac minding the business of their business.

Yet here is what we have today: $640 million worth of bureaucratic busywork and meddling going down the drain for no good reason whatsoever. For crying out loud, the central government should not be funding the $100 million shown below for "consumer education, engagement, and response," whatever that is.

Moreover, banks and financial institutions, which were already the most regulated and supervised businesses in America back in 2010, didn't need another $300 million layer of Washington busybodies and regulatory gumshoes overlooking their activities, as also shown below. They already had the Federal Reserve, Office of the Controller of the Currency, the Federal Deposit Insurance Corporation (FDIC), the SEC, the National Credit Union Administration, the Office of Thrift Supervision, and at least 50 state banking supervisors and regulatory agencies.

In short, termination of the current 1,500 employees of the CFPB and their $240 million annual expense is a no-brainer. So is saving the $400 million balance of the CFPB's budget, which is currently wasted on contractors, grants, advertising, and other superfluous overhead.

And, no, the fact that these expenses are charged to the Federal Reserve's budget is no excuse. Under the law, the Fed remits all system profits to the US Treasury, which profits are now being unnecessarily reduced by the $640 million annual waste of the CFPB.

CFPB Annual Budget

Function	Budget Allocation
Consumer Education and Engagement	$100 million
Research, Markets, and Data	$80 million
Supervision, Enforcement, and Fair Lending	$300 million
Consumer Response and Education	$60 million
Administrative Expenses	$100 million

Agency For International Development (AID)

Foreign aid has always been a waste and failure, even in the context of an Empire First foreign policy and quasi-balanced fiscal environment. But under an America First regime and in fiscal conditions literally hemorrhaging red ink, foreign aid amounts to a Sacred Cow that needs to be slaughtered forthwith.

In Chapter 7 we will amplify on how a true America First national security policy would focus almost exclusively on maintaining an invincible strategic nuclear deterrent and a powerful conventional defense of the coastlines and airspace of North America. But suffice it here to note that wasting money Washington doesn't have on development projects, so-called humanitarian relief, and walking around money for corrupt foreign governments does absolutely nothing for homeland security properly defined.

For want of doubt, here are the 10 largest recipients of U.S. foreign aid (excluding weapons financing) for 2024. Self-evidently, the disastrous proxy war against Russia is enabling Ukraine to absorb the lion's share of the grift. Yet the result of keeping the Ukrainian government on

Washington-infused life support is ultimately a threat, not a boost, to homeland security if it leads to nuclear confrontation with Russia.

Likewise, Ethiopia, Jordan, Somalia, Nigeria, and the now ascendant jihadist opposition forces which overthrew Assad in Syria (who actually get the money) have absolutely nothing to do with safeguarding the liberty of Americans from Maine to Hawaii. And most especially the multi-billion slice of the AID budget which goes to "implementation of the National Strategy for Gender Equity and Equality...(and) to uplift the role of women and girls in all their diversity, including as part of marginalized populations" is purely wokish nonsense utterly irrelevant to national security.

Top Recipients of Non-Military Foreign Aid

1. **Ukraine:** $16.5 billion
2. **Ethiopia:** $2.2 billion
3. **Jordan:** $1.2 billion
4. **Democratic Republic of Congo:** $1 billion
5. **Somalia:** $1 billion
6. **Yemen:** $933.9 million
7. **Nigeria:** $904.4 million
8. **Afghanistan:** $815.1 million
9. **South Sudan:** $794.2 million
10. **Syria:** $748.2 million

Thus, simply closing the Agency for International Development would rid the Swamp of ***10,000*** bureaucrats at an annual direct compensation cost of ***$1.6 billion.*** But that would be just the tip of the iceberg. AID has offices and operations in more than 70 countries around the planet. And these offices are loaded with bureaucrats in the service of Empire First, who are equipped with checkbooks from which upwards of $30 billion of grants are funded annually.

Indeed, zeroing out AID in its entirety is a downright mandatory component of any attempt to slash the Federal budget by $2 trillion. When America is careening toward a ruinous $150 trillion public debt by mid-century it borders on criminal negligence for Washington to send $794 million per year to the likes of South Sudan. The latter is a godforsaken hellhole in central Africa with a GDP of barely $5 billion and per capita income of just $400. Yet AID is shoveling in assistance equal to more than 16% of GDP!

Even more ridiculous is the fact that while Washington has been bombing the Houthi-controlled northern areas of Yemen to smithereens at a cost to the US military of $3 billion in the last three years alone, it is also sending foreign aid of $933 million per year to the government in the south of the country, thereby enabling the Sunni south to pursue its decades-old civil war against the Shiite north. Perhaps it might be more rational to stop both streams of funding and allow the Yemenis to pursue their own civil war in peace–or at least without supervision and meddling by folks on the banks of the Potomac.

And, no, protecting the shipping lanes into the Red Sea is not a matter of US national security. The Chinese container ships and Saudi oil tankers heading for Europe through the Bab-El-Man-deb Strait can always reroute around the Cape of Africa at a small premium if they deem the Red Sea route too hazardous. And, self-evidently, Washington has no business subsidizing cheaper ocean freight to Europe for the oil princes and Chicoms.

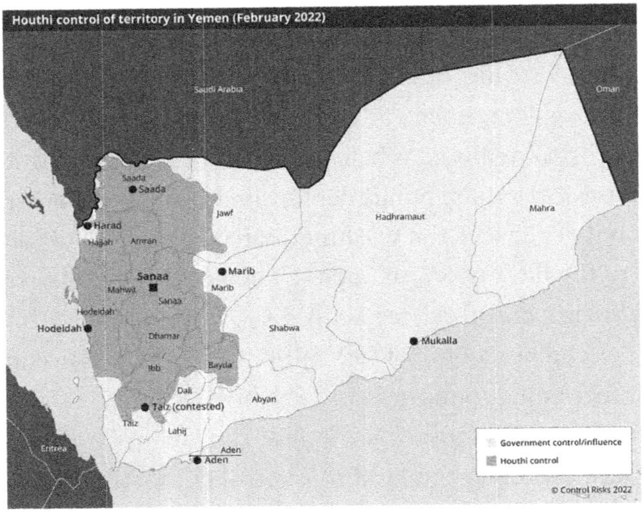

At the end of the day, the Yemen idiocy on display above is no aberration. It represents the inherent stupidity and waste of an imperial foreign policy that attempts to dominate every obscure corner of the planet for no reason of homeland security whatsoever. Therefore, one of the first initiatives of President Trump's pivot to America First must be the complete shutdown of the AID.

FBI

The Federal Bureau of Investigation is a Washington institution steeped in ignominy and disdain for constitutional liberty and democracy. Its forerunner was created during the horrific Red Scare Raids of Attorney General Mitchell in 1919; it flourished prosecuting the idiotic regime of Prohibition during the 1920s; rose to malefic aspect during the Hoover era of Communist witch-hunting and vicious prosecution of civil rights and peace leaders like Martin Luther King Jr; became a fount of false fear-mongering, stings, and entrapment ploys during the War on Terror; and ended up being weaponized by Deep State nomenklatura to destroy the duly elected President of the United States in 2016 and after.

In short, that's 100 years of assault on the rule of law, not its promotion. That history is reason enough to abolish the FBI completely, thereby shrinking the Federal payroll by more than 37,000, at a savings of $6 billion in direct compensation costs and another $5 billion in overhead and operating expenses.

The fact is, there never was a need for the FBI in the first place–outside of political opportunism and the furtherance of crusades which are not within the proper purview of the Federal government. Again, however, we have 90,000 units of state and local government for a reason: That is, to decentralize the exercise of government power, and enforcement of the criminal laws is precisely one of those functions best kept as far away from the nation's capital as possible, as the checkered history of the FBI proves in spades.

In any event, as a practical matter crime prosecution and enforcement is already overwhelmingly conducted by state and local police forces and courts. For instance, there are currently about 7.4 million arrests in the US each year, but only about 10,000 of these are executed by the FBI. That's just ***0.14%.***

Likewise, there are currently ***1,214,000*** police and law enforcement personnel on the payrolls of state and local governments in the US. That compares to just ***15,000*** FBI officers (out of 37,300 staff) involved in domestic criminal law enforcement. This includes all agents and support personnel who work on a wide range of federal crimes such as cybercrime, drug trafficking, violent crime, and white-collar offenses, but, again, it amounts to only ***1.2%*** of the state and local police force level.

At the end of the day, just $2.5 billion of the FBI's $11.4 billion budget is involved in what it generously classifies as "counter-terrorism." We'd say cut that figure by 60% and spin these personnel and activities off to a $1 billion per year counter-terrorism unit in the DOJ. Any real threat of terrorism in the US, as opposed to self-serving FBI concocted stings like the alleged plot to kidnap the governor of Michigan, can be readily handled on a $1 billion annual budget.

After that, close down everything else to the tune of a 34,000-headcount reduction and direct compensation cost savings of ***$5.4 billion*** per year–along with another $5 billion of savings from FBI overhead, contractors, occupancy, travel, and other costs.

Drug Enforcement Agency (DEA)

The War on Drugs was misbegotten from the get-go back when Tricky Dick Nixon launched it in 1970. The only thing it has accomplished is to breed criminals and a brutal underground distribution system funded by the wildly excessive profits owing to the artificial scarcity created by drug law enforcement and interdiction. It has also filled the nation's jails and prisons, mainly on possession charges, thereby providing a taxpayer-funded program where inmates get a free in-house education on how to conduct real criminal activities after their release.

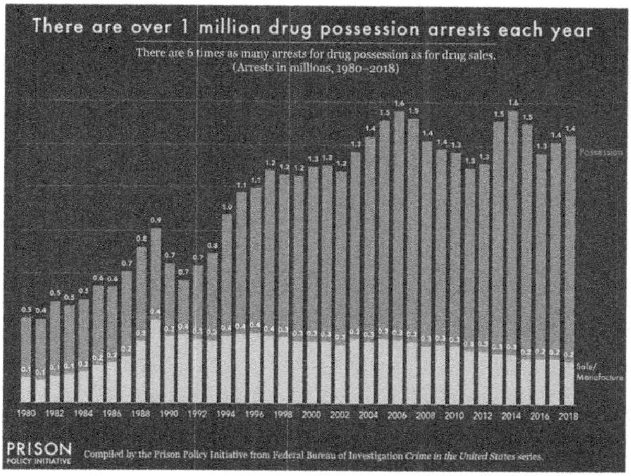

In short, the War on Drugs is a grotesque violation of Market Economics 101. There is simply no other way to characterize the utter stupidity of fostering criminal cartels to do the work of growing, manufacturing, packaging, distribution, and sales that would otherwise be handled by the far more pacific channels of everyday commerce. Indeed, the harsher and more intensive the enforcement against so-called illegal drugs, the greater the amount of crime and the more extensive and tragic the collateral harm that is created as a secondary consequence.

For instance, the plague of fentanyl deaths is clearly owing to the high price of heroin, meth, and other illegal substances stemming from the War on Drugs, which, in turn, encourages the importation and use of fentanyl. Fentanyl is cheaper to produce, easier to smuggle, and extremely potent, making it a lucrative alternative for traffickers. This economic incentive drives its widespread distribution and use, despite its high lethality.

In any event, the surest way to reduce crime at both the borders and in the cities and hinterlands of America alike would be to shut down the DEA cold turkey, releasing 9,300 Federal bureaucrats for more productive work elsewhere. Needless to say, once these hobnailers are off the street, the price of illegal drugs would fall sharply, along with the profitability and incentives for violence among the criminal cartels which run the drug trade.

In all, terminating the DEA would cut its direct compensation cost by $1.5 billion per year, as well as save another $1.6 billion for operations, contractors, and overhead functions. There is hardly any other agency termination candidate where the case is so overwhelming.

Bureau of Alcohol, Tobacco, Firearms and Explosives (ATF)

Ronald Reagan famously said that a government bureau is the nearest thing to eternal life, and surely the ATF is testimony to that aphorism. In 1920 it was born as the Prohibition Bureau, which housed the hated "revenooers." After the Volstead Act was repealed in 1933 it continued to stumble between the Treasury Department and the Justice Department over the decades, looking for missions to justify continued funding. Even in relative bureaucratic obscurity, it gained notoriety for the armed standoff at Ruby Ridge, the extermination of the Branch Davidians at Waco, Texas and the gun "misplacement" scandal of Operation Fast and Furious, among numerous other bureaucratic misfires.

Yet an analysis of what its *5,300* employees and $850 million compensation budget actually accomplish makes clear that the time to terminate the agency arrived long ago. There is absolutely no reason for the Federal government to be in the alcohol, tobacco, and explosives enforcement business at all. Those are inherently functions of state and local government, if they are to be legally regulated and enforced at all.

Allocation of ATF's $1.7 Billion Budget

Category	Budget Allocation
Firearms Enforcement	$500 million
Alcohol Enforcement	$400 million
Tobacco Enforcement	$300 million
Explosives Enforcement	$200 million
Administrative and Other Expenses	$300 million

Likewise, its $500 million budget for "firearms enforcement" is just a polite term for the administration of gun control laws, which self-evidently don't control much of anything. Thus, the number of deaths–both suicide and homicides– due to guns has more than doubled from 20,336 in 1968 to 47,284 in 2021, which translates to rates of ***10.1*** per 100,000 population in 1968 and ***14.1*** per 100,000 in 2021. So much for the ATF's enforcement prowess.

In any event, whatever these ATF bureaucrats might be doing that is necessary and legitimate should be turned over to regular state and local law enforcement. If any rump agency is needed to enforce largely ineffective federal gun control laws–given that there are upwards of one-half billion guns in circulation in the US–these activities can be seconded to a modest residual bureau in the Justice Department at the approximate size of the Office of Violence Against Women ($500 million).

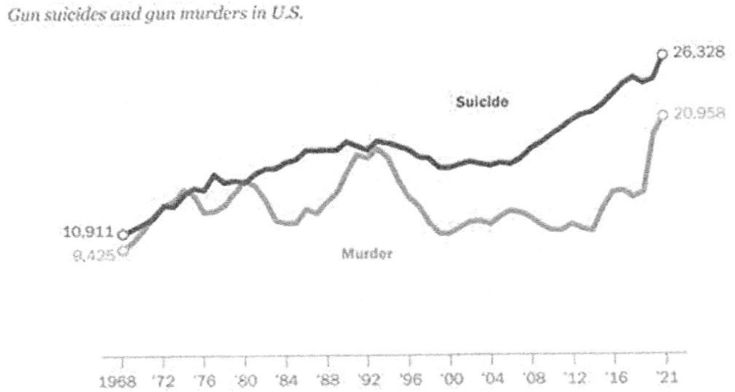

CHAPTER 3

Slash The Fat: Nine Agencies To Cut By 50%

As we have indicated, the 16 agencies slated for elimination do comprise a kind of Litmus Test of fiscal resolve. If these Federal bureaucrats and agencies can't be eliminated, the prospect for reining in America's unfolding fiscal calamity is dim indeed.

Yet the 71,000 headcount reductions and $11 billion of savings constitute hardly a nick out of the Federal payroll. In fact, on a government-wide basis including the uniformed forces, the total payroll numbers nearly ***3.83 million*** civilian and military employees and costs nearly ***$600 billion*** per year. So the agency reductions outlined so far amount to just ***1.8%*** of the total.

For this reason, we have identified a second layer of nine agencies where we believe the headcount could be cut by 50% and existing missions sharply curtailed. This would result in a further Federal staff reduction of 93,000 jobs and nearly $15 billion in direct compensation cost savings.

Agencies DOGE Should Target For 50% Staff Reductions:

- SEC: 2,250 staff and savings of $360 million.
- FCC: 750 staff and savings of $120 million.
- Federal Aviation Administration: 22,500 staff and savings of $3.6 billion.
- IRS: 41,500 staff and savings of $6.64 billion.
- National Labor Relations Board: 800 staff and savings of $130 million.
- Office of Personnel Management: 1,340 staff and savings of $214 million.
- Environmental Protection Agency: 8,500 staff and savings of $1.36 billion.
- NASA: 9,000 staff and savings of $1.44 billion.
- General Services Administration: 6,360 staff and savings of $1.02 billion.
- Total, nine agencies cut by 50%: **93,000** staff and savings of ***$14.9 billion.***

Securities and Exchange Commission (SEC):

The SEC is an exceedingly top-heavy agency with staff of 4,500 absorbing upwards of $1.1 billion per year in compensation expense, which computes to an average cost of $255,000 per head. It is also an egregious Nanny State meddler in an industry–the stock, bond, and related markets of Wall Street–that hardly needs the helping hand of the state to function.

In fact, the basic SEC function–mandated financial disclosure–generates an endless tsunami of filings that are basically superficial, proforma, mechanical, ritualistic, minimally informative, and lawyered ten ways to Sunday.

Consequently, real investors in today's $100 trillion+ Wall Street casino spend lavishly from their own pockets to dig for supplemental business and financial information that can actually make a difference with respect to the prospects and performance of registered companies.

Stated differently, the Wall Street casino is not some kind of latter-day nursery school where the boys and girls operating there need SEC nannies to proof their readers and workbooks. With tens of trillions at stake, investors and traders would get the financial and operational information they need. Failing that, uncooperative or crooked issuers would readily find a torrent of (short) sell orders at the posts where their securities are traded.

SEC efforts to specify and enforce accounting standards are even more of a joke. Virtually all Wall Street stock analysts construct elaborate non-GAAP accounting statements (Generally Accepted Accounting Principles) for the companies they cover and recommend, even as the SEC nannies and gumshoes require companies which mention or discuss these analyst-based non-GAAP versions of their financial results to provide elaborate bridges back to GAAP. So what's the point of spending hundreds of millions per year enforcing GAAP accounting standards when the daily financial vocabulary of Wall Street traders and analysts amounts to a systematic GAAP work-around?

Then comes the foolishness of the SEC's massive market monitoring and enforcement efforts with respect to essentially undefinable and mostly pointless insider trading cases. The very predicate of insider trading laws–that each and every investor should have access to exactly the same information at the same moment in time–is an inherent insult to the basic nature of financial trading markets, where the opposite principle actually pertains. To wit, investors who develop an "edge" are rewarded with superior returns, which is how the market incentivizes and compensates for the search for information and insights that make trading more efficient and productive.

The truth is, insider trading laws amount to a version of flat-earth economics. Besides, the trading moves of a big hedge fund are far more impactful to stock prices than some undisclosed financial tidbit from a company CFO. Yet the latter is per se *illegal* while your "insider" knowledge of trading activity by the big swinging hedge fund operating next door to your own trading floor is completely *legal* so long as it was not obtained through "a breach of duty or trust." Whatever that means.

The argument that without SEC nannies the trading markets would be corrupted by illicitly obtained "inside" information is a relic of what politicians in the early 1930s didn't understand about the real reasons for the 1929 crash (the Fed caused it). But in today's world of instantly and infinitely available information and massive financial incentives to scour the landscape for market moving data, "inside information" amounts to a *Snapchat equivalent.* Its half-life is too short to make a difference to investors over any measurable investment horizon.

The same goes for the SEC's prosecution of "market manipulation." The fact is, there are currently more than 8,000 hedge funds, which collectively manage $2.8 trillion of assets–with much of the latter coursing through the trading desks of a tiny number of "prime brokers" (Goldman, Morgan Stanley, JP Morgan etc.) on a 24/7 basis. These arrangements are fully legal but inherently result in market-moving surges or plunges whenever the big boys all lean in a common direction. By happenstance, of course!

The inherent irregular and sometimes herky-jerky movement of trading markets in response to information and investment flows generated by the hedge fund mob, in fact, is just Economics 101. By comparison, the SEC's market manipulation targets amount to chasing the ghosts of what sharp-edged traders did in a more primitive time way back in the 1920s. In today's technology-enabled trading world none of these alleged marketplace sins would actually have more than a momentary impact and, in any event, would breed countermanding schemes similar to the manner in which some traders today buy excessively shorted stocks to trigger a covering rally. That is to say, today's markets root out cheaters and short-cutters far faster than could any passel of overpaid GS-16s at the SEC.

In short, the Trading Nannies at the SEC are not needed to prevent any of the five types of "market manipulation" pursued by the agency. Today's information, communications, and technology-rich Wall Street markets, where trillions are at stake every minute and hour, are always and everywhere on the alert for so-called abuses of these types. They don't need the SEC to tell them when marketplace miscreants are pumping, dumping, spoofing, wash trading, churning, or lying. It quickly becomes obvious to other traders who are paid to stay alert.

1. **Pump and Dump Schemes:** Inflating the price of a stock through false or misleading information, then selling it at the inflated price.

2. **Spoofing:** Placing fake orders to create a false impression of supply or demand, influencing other market participants.
3. **Wash Trading:** Simultaneously buying and selling the same security to create the illusion of increased trading volume.
4. **Churning:** Excessive trading by a broker in a client's account to generate commissions without regard for the client's investment objectives.
5. **False Statements and Rumors:** Spreading false information to manipulate stock prices.

At the end of the day, the SEC should be retired to the museum of 1930s mythologies. Yet it is probably so deeply embedded in the warp and woof of the financial markets–especially with respect to routine reporting in 10Qs, 10Ks, offering prospectuses, etc.–as to make its complete abolition impractical. Still, it is involved in so much unadulterated Nanny State nonsense and tomfoolery that its massive staff and payroll could be easily cut in half.

That would reduce the Federal headcount by ***2,250*** bureaucrats and upwards of ***$575 million*** of annual expense. The fact is, the SEC *modus operandi* itself is an exercise in government-conducted fraud, waste, and abuse because its Nanny State regime doesn't make investing on Wall Street any safer– just more expensive and cumbersome. So if 50% of that kind of "fat" can't be cut, there is not much hope of achieving the much more difficult savings from downsizing the muscle or cutting the bone in the rest of the Federal budget.

Federal Communications Commission (FCC)

The FCC is another relic of 1934. Even a brief perusal of the manner in which it allocates its $411 million annual budget tells you that much. In an internet-based streaming, satellite-TV, and Starlink-enabled world there is no longer any such thing as oppressive communications monopolies. Competition is exceedingly fierce, innovative, and far-reaching–as perhaps crystalized by the "cable-cutting" crisis now sweeping the traditional cable TV business.

As it happens, traditional cable TV has seen a cumulative loss of **over 10 million** subscribers just since 2018, and the loss has been accelerating, with major providers like Comcast and Charter reporting relentless declines. So why in the world are the busybodies at the FCC in the business of consumer protection, competition promotion, or broadband extension to so-called underserved areas, at all? Technological advance and competitive innovation have all these matters covered in spades without any help from the FCC nannies seeking to justify their jobs.

Indeed, the FCC nannies are mostly a hindrance to pro-consumer innovation, as is evident in the rigid blunderbuss of the "net neutrality" rules. Again, this is just regulatory socialism which prohibits internet service providers (ISPs) from blocking, throttling, or engaging in paid prioritization of any lawful content, such as providing premium-priced "fast lanes" for streaming services.

That is to say, these bureaucratic power-grabbers through a new order as of April 2024 are attempting to shoehorn today's world of internet and technology-based dynamism into the placid one-size-fits-all Ma Bell modality of 1970.

At the end of the day, the only arguably useful thing the FCC does is public spectrum management, but even that should be limited to auctioning licenses for different categories of media–such as television, radio, mobile communications, and public safety. And that would require only a tiny fraction of today's staff and budget.

In short, the FCC would be getting off easy with a 50% staff cut. That would reduce its headcount by 750 jobs, save direct compensation costs of at least $120 million per year, and would also enable much of the remainder of its budget (@$300 million) to be eliminated, as well.

Breakdown of the FCC's $501 million budget for fiscal year 2024:

Category	Budget Allocation
Spectrum Management	$150 million
Consumer Protection	$80 million
Public Safety and Homeland Security	$70 million
Promoting Competition	$50 million
Broadband Deployment	$100 million
Enforcement	$51 million

Federal Aviation Administration (FAA)

The FAA gives the notion of a padded-payroll bureaucracy an altogether new definition. Its 45,000 staff positions exceed by orders of magnitude the levels that should be needed to efficiently accomplish the air safety and traffic control functions it has been assigned by the Congress.

For want of doubt, just consider the enormous bloat in the

20,000 staff level at the Air Traffic Control Organization (ATO). A few years back the DOT inspector general (IG) performed a detailed comparison of costs and staffing levels as between–

- The 254 airport control towers that are operated and staffed by *outside contractors*, with ***1,400*** controllers who manage 28 percent of the National Airspace System.
- The 266 airport towers that are operated directly by the FAA, with ***15,000*** controllers who manage the other 72 percent of the system.

The IG found that contract towers were roughly ***three times more cost effective*** per aircraft movement handled than FAA towers. Yet the contract tower safety records for comparably-sized facilities were also on par with those of FAA towers.

On average, the report said, contract towers use 48% fewer resources per aircraft handled per year. That's the case even though economies of scale should favor the larger FAA-run towers which handle significantly more flights.

Moreover, the IG found that the contract towers deliver lower costs because they typically staff their facilities with fewer controllers whom they pay much less than the FAA pays its staff. The report said a year's worth of labor and benefit costs at the busiest FAA towers runs about $15.7 million compared to just $2.7 million for a contract facility.

Moreover, standardizing these figures on a *per aircraft movement* basis showed that the FAA facilities cost ***$22.34*** per unit compared to ***$7.41*** per unit at a contract tower. Further evidence for the gross inefficiency of the ***FAA-run tower***s lies in the fact that the cost per aircraft movement figure remained *relatively constant* as between busy and less busy airports, thereby revealing formulaic staffing rather than the scale-based efficiencies that should be realized with increasing volume levels. By contrast, the latter were very much in evidence in the ***contract towers***, where per movement costs increased by roughly 30 percent at the slowest, lower-volume towers.

In short there is every reason to believe that the 15,000 controllers and 20,000 total staff at the FAA-run towers could be readily cut by 50% if they were operated by efficient outside contractors.

Distribution of the FAA's 45,000 staff among the major functions:

Category	Staff Allocation
Air Traffic Organization (ATO)	20,000
Aviation Safety (AVS)	5,000
Commercial Space Transportation (AST)	1,000
Facilities & Equipment (F&E)	4,000
Research, Engineering & Development (RE&D)	3,000
Grants-in-Aid for Airports (AIP)	2,000
Security & Hazardous Materials Safety (ASH)	1,500
Finance & Management (AFN)	1,500
NextGen (ANG)	2,000

The bloat is more than evident in the staffing levels for the other functions, as well. For instance, the FAA's 1,500 staffers in just the finance and management area compare to only 100 or so employees in the legal and finance functions at X; and the 9,000 employees in the RE&D, Facilities, and Equipment and NextGen technology modernization functions compare to about 800 total employees in engineering and production functions at X.

Likewise, the 1,000 staff in the Commercial Space Transportation division essentially oversee just 50 launch service providers including, notably, SpaceX and Blue Origin and numerous far smaller, specialized operations; and, also, about 10 spaceport operators. So the level of staff padding in this department is self-evidently massive.

But the real staffing jackpot lies in the AIP function which funds airport infrastructure projects such as runways, taxiways, airport signage, airport lighting, and airport markings. As is evident in the chart above, the FAA employs 2,000 staffers to bureaucratically stir, blend, mix, knead, and return to local airports the ticket tax and landing fee revenues that should never have been roundtripped to Washington in the first place. That is, adopt a policy of eat what you kill in terms of airport fees and capital investment, and all 2,000 AIO bureaucrats would be redundant.

In short, cutting the FAA staff by 50% should be very easy. The reduction of 22,500 headcounts would save in the order of $3.6 billion per year, with collateral savings from reduced overhead and support costs of roughly equal magnitude.

National Labor Relations Board (NLRB)

The NLRB is another relic of 1935 that is barely needed to enforce the nation's now well embedded and legally evolved labor laws. And like most other agencies on the banks of the Potomac, it has been captured by the unions it supposedly regulates.

After 89 years, in fact, the rights and protections of the National Labor Relations Act (NLRA) are well-settled law and deeply institutionalized in practice. Specifically, the NLRA and its case law protects the rights of employees to organize, form, join, or assist labor organizations, to bargain collectively through representatives of their own choosing, and to engage in other concerted activities for the purpose of collective bargaining or other mutual aid or protection. It also prohibits employers from interfering with, restraining, or coercing employees in the exercise of these rights.

All of these rights, protections, and safeguards, therefore, could be readily enforced by the Federal court system in lieu of the NLRB bureaucracy. The latter's separate and parallel administrative and adjudicatory system is simply regulatory capture at work. The unions like to own their own judges.

To completely shift NLRA enforcement to the regular Federal court system would likely take statutory amendments. But by proposing to cut the NLRB staff by 50%, thereby reducing Federal headcounts by 800 and compensation costs by $130 million, DOGE would pave the way toward the long-overdue displacement of an obsolete agency that survives mainly for the service of its political clients, not the public good.

Internal Revenue Service (IRS)

Needless to say, we are not talking about just bureaucratic nannies and meddlers in the case of the current ***83,000*** IRS employees--a figure which is heading for 102,000 by the end of the decade under the Biden revenue grab. In the ranks of what amounts to a small city's worth of Federal bureaucrats are also a goodly phalanx of tax cops, gumshoes, enforcement lawyers, and tax filing proctologists.

So the question recurs: What has generated this massive bureaucracy in the first place, and what fundamental policy shifts are needed to cut the IRS headcount by 50% (42,000 jobs) and upwards of $5 billion of compensation and other operating costs?

The answer starts with calling the IRS' bluff. When you look at the actual tax filing data it is damn evident that the Deep State bureaucrats are faking mightily when it comes to their massive staffing demands. We discovered the scam way back in our OMB days while jousting with the Treasury Department over the sacred cows in its budget. But nothing is different 40 years later–so here's the smoking gun that points the way.

In the most recent complete tax-year (2022) there were 161.336 million individual income tax returns filed, which reported $14.83 trillion of Adjusted Gross Income (AGI). But fully ***146.045 million*** of those filings, which reported $10.025 trillion in AGI, ***did not claim any itemized deductions.***

Moreover, among these non-itemizers about ***97 million*** owed taxes and used the standard deduction to calculate taxable income and the amount owed before tax credits. And another ***49 million*** standard deduction users owed no Federal income taxes at all due to low taxable income or child and other tax credits.

In short, you absolutely do not need a giant IRS bureaucracy–and, indeed, hardly any labor-intensive operation at all–to administer the IRS code in the case of ***92%*** of annual tax filings and the overwhelming share of US income taxpayers. That's because virtually all the relevant data for completion of these non-itemized filings is machine-readable and available on other IRS reporting systems, as shown below.

For want of doubt, here is the entirety of the tax computation for a couple earning wages at the US median income of $80,000, using the $13,850 standard deduction for a joint return and claiming two $2,000 child tax credits. The fact is, with today's technology 99.999% of the work of processing, examining, and adjusting (if necessary) non-itemized tax returns of this type should be accomplishable by IRS computers, with nary a bureaucrat's fingerprints evident in the whole shebang.

That's especially the case because the overwhelming share of the $10 trillion of AGI among non-itemizers is for wages and salaries reported to the IRS on W-2s; and, also, for interest, dividends, rents, royalties, independent contractor earnings, stock sales, pensions, annuities, IRAs, and taxable Social Security earnings–all of which are also reported by the payers of these amounts on Form 1099s.

In the illustration below, any alert machine–to say nothing of an AI-enabled one rigged-up by Elon & Co.–could cross-check the W-2s, calculate the taxable income, apply the three relevant tax brackets, deduct the $4,000 of child credits, verify the tax liability of $6,243 and subtract the taxpayer's withholding amounts to determine whether a payment or refund was required. All in literally a nanosecond. Presto!

Step-by-Step Calculation of $80,000 Wage-Earning Couple With 2 Kids and The Standard Deduction

1. **Gross Income:** $80,000
2. **Standard Deduction:** $13,850 (for married filing jointly in 2023)
3. **Taxable Income:** $80,000 - $13,850 = $66,150

Tax Liability Calculation

1. Taxable Income: $66,150
2. Tax Rates:
   - » 10% on income up to $22,000
   - » 12% on income from $22,001 to $38,600
   - » 22% on income from $38,601 to $66,150

Tax Calculation

- **10% on the first $22,000: $22,000 * 0.10 = $2,200**
- **12% on the next $16,600: $16,600 * 0.12 = $1,992**
- **22% on the remaining $27,550: $27,550 * 0.22 = $6,061**

Total Tax Before Credits: $2,200 + $1,992 + $6,051 = $10,243

Child Tax Credits

- 2 Child Tax Credits: $2,000 per child
- Total Credits: 2 * $2,000 = $4,000

Tax Liability After Credits: $10,243 - $4,000 = $6,243

So how in the world can they justify **83,000 bureaucrats** and a budget of $16.1 billion when the overwhelming share of returns involve what can be aptly described as "Machine Work 101?"

The answer, purportedly, lies in the balance of filings–the ***15.29 million*** itemized returns. But even here the overwhelming bulk of the relevant income items and deductible expense items are not so complex or opaque at all. Indeed, they too are available on other IRS reporting systems and are machine-readable at the individual taxpayer ID level.

Specifically, in 2022 the amount of AGI reported on these itemized returns was ***$4.809 trillion*** or about 32% of total AGI. But within this total there was included the following amounts which are all *machine-readable* from W-2s and 1099s:

- Wages and salaries: $2,345 billion.
- Taxable interest: $78 billion.
- Taxable dividends: $230 billion.
- IRAs: $110 billion.
- Pensions and annuities: $176 billion.
- Taxable Social Security $82 billion.
- Unemployment benefits: $3 billion.
- Subtotal: ***$3.024 trillion.***
- Above Income As Percentage of AGI on Itemized Returns: *63%.*

When it comes to verification on a machine-readable basis, the above income items are all check, check, and check. This means that when you combine the above machine-readable AGI amounts from itemized returns with the $10.025 trillion reported on non-itemized returns, it works out to ***88% or $13.049 trillion*** of the $14.834 trillion of total AGI reported for 2022.

None of this AGI should require any significant labor-intensive administration, examination, adjustment, or enforcement. The IRS computers should be aware of every dime of AGI from the above categories and whether it was filed accurately by the taxpayer or in need of the proverbial IRS-ordered "adjustment."

So it is hard to figure out why on the AGI/income side of the equation there is a need for anything remotely like the headcount and budget magnitudes shown below. For instance, the *22,000* headcounts in "enforcement" and "compliance" should be as idle as the proverbial Maytag repairman when it comes to standard deduction returns and the machine-readable sources of income filed on itemized returns. In these instances, there is nothing material for taxpayers to cheat on that wouldn't be flagged by a properly programmed computer instantly upon filing.

And the same is true in the case of the *33,000* headcounts in "taxpayer services," "operations support," and "administrative support." Virtually none of these bureaucrats are needed to process the $10 trillion of AGI on the 146 million non-itemized returns and the $3 trillion of AGI on 15 million itemized returns that is already reported independently by the underlying payers of these income sources.

To the contrary, that's work for 24/7 machines, not 6.5 hours per day (after civil service required breaks and lunch) government bureaucrats who get 35 vacation, personal leave and sick days per year, on average. And a high share of whom in the post-pandemic era don't deign to come into the office even on workdays, anyway!

Breakdown of IRS Budget and Headcount By Function:

Function	Budget (in billions)	Headcount (FTE)
Taxpayer Services	$2.78	8,000
Enforcement	$5.44	7,000
Operations Support	$4.10	10,000
Business Systems Modernization	$0.29	1,990
Administrative Support	$1.50	15,000
Compliance	$1.50	15,000
Other Functions	$1.00	2,010
Total	$16.61	83,000

For want of doubt, here are the other arguably more complicated categories of AGI reported on itemized returns. But even in these cases, there is plenty of work for the machines to do with respect to examination and verification. For instance, *$845 billion or nearly half* of the total below is owing to capital gains. But that source of income is already reported on Form 8940 and Schedule D of 1040s. So Elon's machines should be on top of that, as well.

At the end of the day, most of the complexity and opacity of the IRS code relates to the $106 billion of net profits reported for business and professional income and the $704 billion reported by subchapter S corporations. Here the complexity arises not just from gross income reporting, but more especially from the timing and amounts of allowable business expenses incurred in getting to the net profits figures shown below.

Still, the total amount of AGI involved in these two sources at $950 billion is just ***6.5%*** of total AGI. Even if returns with heavy subchapter S or professional and business earnings involve a lot of digging, checking, reconciling, and verifying by humans, it is hardly likely to be anything close to 83,000 bureaucrats' worth.

So, yes, there may well be 2,650 pages of IRS code and another 9,000 pages of IRS regulations, and the whole thing may have 25 times more words than the *Lord of the Rings* trilogy. But when it comes to the overwhelming bulk of income tax filings and the AGI reported on them, 98% of this legal labyrinth is largely irrelevant. Its unfortunate existence is merely cited as a smokescreen to justify a massive, unnecessary tax collection bureaucracy.

Other Sources of Income Reported on Itemized Returns in 2022

- Business and professional profits: ***$106 billion.***
- S-corporation net income: ***$704 billion.***
- Capital Gains: $845 billion.
- Property sales: $26 billion.
- Rents and royalties: $20 billion.
- Estate and trust income: $29 billion.
- Gambling income: $29 billion.
- Other, net: $5 billion.
- Total of above: ***$1.764 trillion.***

Nor does the itemized deduction side of the ledger change the picture. Upwards of *91%* of itemized deductions, which amounted to *$669 billion in 2022*, were accounted for by the first five line-items shown below. These are largely machine-readable based on standard reporting forms that originators of these deductions are required to file with the IRS.

For instance, mortgage interest deductions are reported on

Form 1098; charitable contributions are reported in Form 990 and deductions for state and local taxes paid are available from IRS information sharing reports by the states. Yet in 2022 these three deductions alone amounted to nearly $500 billion or 75% of the total.

Major machine-readable itemized deductions in 2022:

- Medical deductions after floor: $93 billion (gross deductions of $121 billion less $28 billion floor effect).
- Taxes paid deduction: $125 billion.
- Mortgage interest deduction: $147 billion.
- Investment interest deduction: $23 billion.
- Charitable contributions deduction: $222 billion.
- All other itemized deductions: $59 billion.
- Total Itemized Deductions: ***$669 billion.***

In short, upwards of ***90%*** of the AGI reported in 2022 for all returns was machine-readable from independent reporting sources and more than ***90%*** of itemized deductions were also machine-readable. *Accordingly, the preponderant share of income and deduction data coursing through 161 million annual income tax filings is essentially riding in a self-driving vehicle.* The work of processing, assessing, validating, and adjusting it, where necessary, does not likely require more staff than the current headcount of the Capitol Hill Police (2,400).

Moreover, even a few small intelligent changes in the IRS code would narrow even further the number of returns and amounts of AGI that need labor-intensive review and verification. For instance, among the 15.29 million itemized returns filed in 2022, the overwhelming share were at the lower and moderate ends of the income scale where disputed deduction amounts are inherently limited.

2022 Distribution of Itemized Deduction Returns By AGI Level:

- $100,000 and under: 5.755 million (37.6%).
- $100,000 to $500,000: 8.076 million (52.8%).
- $500,000 to $1,000,000: 0.903 million (5.9%).
- $1,000,000 and over: 0.558 million (3.7%).
- Total Itemized Returns: 15.292 million (100%).

As indicated above, 13.831 million itemized returns, or more than 90%, reported AGI of $500,000 or lower--including 5.8 million at $100,000 or lower.

In turn, these $500,000 and under filings reported an aggregate of $1.960 trillion of AGI and $475 billion of itemized deductions. So a "variable standard deduction" allowance of roughly this ratio–24% of AGI– for currently itemized returns up to $500,000 would be *revenue-neutral.* But by eliminating upwards of 90% of itemized deduction filings, an income-based "variable standard deduction" would also surely reduce the need for several thousands of examiners, service personnel, and overhead managers, as well.

After all, there are only ***1.461 million*** returns with AGI of $500,000 or higher, reported in the amounts shown below for 2022. We absolutely do not believe, for instance, that you need a bureaucracy of 83,000 to examine ***$213 billion*** of itemized deductions taken by the wealthy, when ***$126 billion*** of these were owing to charitable contributions and ***$41 billion*** to investment interest deductions. That's nearly 80% of the total deductions taken by the wealthy, yet every dollar of this is machine-readable and verifiable because it is reported independently to the IRS by the charitable institution recipients and interest-receiving banks, respectively.

Likewise, 71% of the $2.78 trillion of AGI is due to W-2 salaries ($931 billion), investment interest ($62 billion), ordinary dividends ($170 billion), capital gains ($774 billion), and rents and royalties ($40 billion). As indicated previously, all of these items are also reported to the IRS separately and are machine-readable by its computers at the taxpayer ID level.

So even in the insane nest of complexity which is the US tax code as it applies to the wealthy, the case just isn't there to justify the egregiously padded payrolls at the IRS. Not even remotely–and that's before taking a legislative scalpel to the tax code with the aim of drastically broadening the base and flattening the rates.

Key Tax Data for the 1.461 Million Filings with AGI of $500,000 or Higher:

- Total AGI: $2,780 billion.
- Itemized Deductions: $213 billion.
- Other Adjustments: $86 billion.
- Taxable Income: $2,481 billion.
- Taxes Paid: $708 billion.
- Itemized Deductions as % of AGI: 7.6%.
- Taxes Paid As Percentage of Taxable Income: *28.5%*.

So can DOGE find a way to cut the IRS staff by upwards of 50% and 42,000 bureaucrats at a budget savings of $5 billion per year? We'd say, yes, just dig into the rich trove of data in the IRS Data Book for recent tax years and the degree of the current scam will become more than evident.

In that context, DOGE might well consider a technologically modern version of the old postcard-based approach to simplification of the Federal income tax. Thus, there is no reason why upwards of 150 million filers with AGIs under $500,000 could not simply receive an "E-Card" from the IRS at their personal email address in which the IRS machines have already done all the work. The E-Card would:

- Calculate and sum all sources of AGI.
- Apply the standard deduction and child credits.
- Compute the tax liability owed.
- Calculate the amount of either payment or refund due after crediting taxpayer withholdings.
- Provide an option to "accept" the E-Card outcome or elect to file different amounts in the regular way.

Again, based on the IRS filing data it is likely that at least 90% of E-Card recipients would check the "accept" box and be done with tax time, with no expense on their end and no IRS bureaucrats on the other end.

Our confidence in this conclusion is based on these considerations from the 2022 IRS data. Only 4.7% of the 120 million returns with AGI under $100,000 claimed itemized deductions and the amount of AGI on these returns was just 7% of AGI on all returns under $100,000.

Summary of Filings with AGI of $100,000 or Under:

- Returns With No Taxes owed: 47.048 million returns with $922 billion of AGI.
- Standard Deduction Returns: 66.865 million returns and $3.424 trillion of AGI.
- Itemized Deduction Filers: 5.755 million returns with $345 billion of AGI:

Likewise, 1.88 million or 22% of filings in the $100,000 to $500,000 range used itemization, but it is likely that the aforementioned variable standard deduction approach would limit the number of itemizers very sharply.

Finally, even without sweeping tax reform and the substitution of tariff and consumption taxes for the income tax, as has been proposed by President Trump, a huge share of the $550 billion that taxpayers now absorb to figure and file their taxes and contest with the IRS bureaucracy could be readily and substantially reduced.

Office Of Personnel Management (OPM)

This entire agency is essentially a centralized HR operation for the 2.2 million civilian labor force (FTE basis) of the Federal government. But it is truly redundant because the big departments and agencies maintain their own HR operations, which in theory are geared to the specific missions and operating conditions of each agency.

For instance, the Veterans Affairs Department, with a total payroll of 412,000 employees has 19,000 HR staff. That's one for every 22 employees. Likewise, EPA has 583 staff positions and $402 million of budget devoted to "HR Solutions." That amounts to one HR staffer for every 29 EPA employees. So how could these deeply padded client agency operations need even more HR support help from OPM?

Similarly, the Homeland Security Department has 1,000 HR staff in its Office of the Chief Human Capital Officer (OCHCO), while the DOJ has about 1,000 HR staff in its Justice Management Division (JMD). Most of the other large agencies including HHS, Treasury, USDA, and the Interior Department also have large-scale internal HR departments that number upwards of 500 to 1,000.

Self-evidently, the theory of outsourcing HR-type service functions to a central bureau was borrowed from the corporate world, but this kind of outsourcing does not work in the labyrinthine arena of the Federal government. That's especially the case given that agency operating budgets are controlled by 12 different appropriations committees which often have vastly different agendas. Accordingly, all the HR and related "servicing" operations that have been centralized at the OPM should be assumed to be duplicative of those that already exist at the departmental and independent agency level.

At minimum, therefore, the 583 headcounts in OPM's HR Solutions division should be eliminated, as well the 18-person DEI staff and a goodly share of the nearly 1,400 staff in the health insurance and retirement services operations. All the large agencies have their own specialists to assist employees with health insurance and retirement matters.

Likewise, the 121 staffers and a $140 million budget for the Chief Financial Officer function, which amounts to 12% of OPM's budget, is beyond ridiculous. That's because the client departments and agencies which OPM services already have their own financial control offices and functions. So the only reason for this office in OPM is internal financial control, which could be handled with a desktop computer and a small staff numbering 10 employees or less.

In fact, the only valid function for OPM is the original idea that it would set *policy* and employment *standards* for the other agencies, not operate as a far-flung employee support services department for the entire Federal government. A thorough downsizing program based on this core principle would easily support a 50% reduction in the budgets and headcounts shown below for the current OPM offices.

OPM Budget and Staff Levels by Major Functions (FY 2024)

Major Function	Budget (in millions)	Staff (Total Headcount)
HR Solutions	$402.2	583
Healthcare & Insurance	$78.6	257
Retirement Services	$152.0	1,124
Facilities, Security & Emergency Management	$17.3	83
Office of the Chief Information Officer	$238.2	305
Office of the Chief Financial Officer	$140.2	121
Office of the Inspector General	$45.8	212
Office of Communications	$3.4	18
Office of Procurement Operations	$7.9	34
Office of the General Counsel	$8.7	40
Office of Diversity, Equity, Inclusion & Accessibility	$6.5	18
Office of the Executive Secretariat	$8.3	33
Human Capital Data Management & Modernization	$68.4	68
Office of Privacy and Information Management	$0.0	0
Other Functions	**$22.2**	**323**
Subtotal, All Functions	**$1,177.7**	**3,219**
Adjustment to FTE Basis	**N/A**	**-719**
Total (FTEs)	**$1,200.0**	**2,500**

Environmental Protection Agency (EPA)

The key to downsizing the bureaucratic monstrosity otherwise known as the EPA is straightforward. To wit, all the woke and left-wing ideological digressions which absorb enormous budgetary resources and manpower but have nothing to do with efficient minimization of legitimate pollution externalities in the areas of air, water, land, and solid wastes should be eliminated.

For instance, the current 2022-2026 EPA strategic plan focuses on seven goals but the first two have nothing whatsoever to do with it statutory missions:

1. "**Tackle the Climate Crisis:** Reduce emissions, accelerate resilience, and advance international climate efforts."
2. "**Advance Environmental Justice and Civil Rights:** Promote environmental justice at federal, tribal, state, and local levels."

Thus, DOGE should root out every staffer and budget dollar devoted to DEI and to so-called greenhouse gas emissions and the regulatory rule-makings and enforcement activities predicated upon net zero. As a matter of science, CO2 is not even a pollutant but rather is fertilizer to the very life of the planet–even as these regulatory misfires waste EPA resources and shackle production, employment, and economic growth in the private sector.

At the operational level, this right-sizing to its legitimate statutory functions would result in enormous savings in staff and budget outlays. For instance, here is the EPA's own summary of its current activities on the Climate Crisis which have no defensible statutory basis:

EPA's Efforts Against the Climate Crisis

1. **Measuring Emissions:** The EPA tracks greenhouse gas emissions using programs like the **Inventory of U.S. Greenhouse Gas Emissions and Sinks** and the **Greenhouse Gas Reporting Program.**
2. **Reducing Emissions:** The EPA works with industries to reduce greenhouse gas emissions through regulatory initiatives and partnership programs.
3. **Climate Adaptation Plan:** The EPA's **2024-2027 Climate Adaptation Plan** focuses on building climate resilience and preparing for climate impacts.
4. **Advancing Science:** The EPA conducts research to understand climate change impacts and develop strategies to manage risks.
5. **International Partnerships:** The EPA collaborates with other countries to address global climate change.

Nearly all of the above should be eliminated under a strict interpretation of the EPA's legal mandate as recently clarified by the Federal courts. The power-seeking bureaucracy at EPA has long insisted that its authority to regulate greenhouse gases arises from the Clean Air Act (CAA), which, unfortunately, the Supreme Court initially greenlighted in *Massachusetts v. EPA (2007).*

However, the court has subsequently clarified in *West Virginia v. EPA* (2022) that the CAA does not grant the EPA such expansive regulatory powers over greenhouse gases and that it may not implement broad-based controls on them. Accordingly, the EPA's current regulatory dragnet needs to be purged of all rules, regulations, enforcement, and other administrative activities that run afoul of a strict interpretation of the *West Virginia* ruling.

That's especially warranted because the Biden Administration has made every effort to circumvent the court's decision. For example, the EPA has proposed new carbon control rules that focus on reducing emissions at individual facilities rather than implementing its previous power generation-shifting measures on a broad regional and national basis, falsely claiming that this ruse is compliant with *West Virginia*. But it surely is not.

Nor is the area of greenhouse gases the only place where regulatory overreach needs to be drastically curtailed. For instance, under the Clean Water Act the EPA has gone far astray in interpreting "waters of the United States" (WOTUS) to include thousands of isolated, temporary and seasonal waters, wetlands, ponds, and puddles that go way beyond any rational case for Federal protection.

Indeed, the latitude for a drastic scaleback of EPA regulatory and enforcement activities based on strict adherence to explicit Congressional intent and the black letter of the law is enormous. For instance, in 2020 alone the EPA published more than 1,400 documents, rules, and orders in the Federal Register that could not possibly have been within the explicit intent of the statutes cited.

Obviously, a strict constructionist purging of the EPA's regulatory and enforcement dragnet would enable a substantial reduction in its 17,000 headcount and the related estimated compensation cost of $3 billion. Yet beyond that, it is abundantly clear that the *EPA is a world champion of padded payrolls*.

For instance, only about 9,500 of its 17,000 FTEs are attributable to its core function of "environmental programs and management." All of the remainder are in science and technology research, liaison with state and tribal operations, and various operational roles involving Superfund sites, leaking storage tanks, oil spills, etc. A large share of these activities are either redundant, unnecessary, or could be contracted out by competitive bid at far lower cost.

In short, it would be virtually a no-brainer to cut 8,500 positions from the EPA payroll at an estimated savings of $1.4 billion per year in fully-loaded compensation costs. And that would be just the tip of the iceberg.

NASA

NASA's current budget is about ***$25 billion,*** and its staff count is ***18,000.*** However, the implied payroll cost of $2.88 billion at our fully-loaded assumption of $160,000 per head by FY 2029 accounts for only 12% of the total. The preponderant budget costs at NASA, therefore, are for *outside contractors and vendors of goods and services.*

These costs totaled about ***$19.6 billion*** in FY 2023 and break out as shown below. Approximately $3.8 billion of these costs, mostly in the first line item, were paid to SpaceX for various services such as rocket launches, cargo missions to the International Space Station (ISS), and development work for NASA's Artemis program.

Breakdown of NASA's $19.6 billion spent on outside contractors and vendors for fiscal year 2023:

Category	Estimated Spending
Aerospace and Defense Contractors	$8.0 billion
Information Technology Services	$2.5 billion
Engineering and Technical Services	$3.0 billion
Construction and Facilities Management	$1.5 billion
Research and Development	$2.0 billion
Professional Services	$1.0 billion
Other Services	$1.6 billion

Needless to say, Elon Musk would know exactly how much padding and excess is built into these contracts, which figure is undoubtedly massive and amounts to billions per year. And he would also surely recognize that NASA's entire budget in the context of the current public debt explosion is *entirely discretionary*: A nation heading for fiscal calamity simply doesn't need anything that NASA provides–and certainly not if it's funded by piling more debt on future taxpayers.

So NASA should consider itself lucky to be assigned a 50% across-the-board cut for all of its outside contracts and in-house staff costs alike.

In the latter case, that would amount to a 9,000-headcount reduction, with an estimated compensation cost savings of $1.44 billion. In addition, contractor savings would amount to another $10 billion per year, as outlined below.

Obviously, a 50% budget cut would have to be allocated among the 10 major program areas currently being pursued by NASA. While there are no credible national security or near-term societal benefits from any of these programs–*they amount to a swell taxpayer-funded adventure for humanity*–it is possible to distinguish between programs that have more of a pure science than a commercial aspect and reserve the funding priority for the former.

That is to say, let the billionaires and venture investors absorb more of the cost for the Artemis, Space Launch, and Moon programs while reserving funds in the residual NASA budget for the Space Telescope, astrophysics, and planetary science programs. In any event, $12.5 billion and 9,000 staffers is surely the most the nation can afford in the face of public debt heading toward $150 trillion before mid-century by CBO's own reckoning, and probably well before the royalties from any ballyhooed Moon and Mars mining programs could make any difference.

Breakdown of NASA›s Major Programs and Activities In 2023

Program	Budget (in billions)	FTEs
Artemis	$7.48	2,000
James Webb Telescope	$0.6	500
Space Launch Systems (SLS)	$2.5	1,200
Orion Multi-Purpose Crew Vehicle	$1.2	800
Exploration Ground Systems	$1.0	700
Earth Sciences	$2.2	1,500
Planetary Sciences	$3.2	1,200
Astrophysics	$1.5	1,000
Aeronautics	$0.8	800
STEM Engagements	$0.15	200
All Other Functions	$4.0	1,000
Total	**$25.0**	**18,000**

General Services Administration

Slashing the $11 billion budget and 13,000 staff positions at the GSA would be a no-brainer in the context of a $2 trillion budget cut program that includes a *40% reduction in the nondefense payroll.* After all, huge portions of the 360 million square feet of office and other space currently under GSA management, which is contained in 8,400 buildings spread across 2,200 communities from sea to shining sea, would become vacant and redundant.

In that context, the first to get pink slips should be the 3,000 staffers at GSA's Public Buildings Service (PBS) division. The latter's mission is "to construct, manage, and preserve government buildings." Given that millions of square feet of GSA space will become vacant the last thing the agency should do is spend one more dime on the location, acquisition, construction, or fitting out of additional government-owned office space.

In fact, of the 690 *major* buildings managed by the GSA about 500 are leased and just 190 are owned. Since it is locked into legally binding long-term leases on the first 500 buildings, the obvious thing for GSA to do after the "Fat is Slashed" from the Federal workforce is to sell the owned buildings and consolidate Federal activities in the leased buildings–an action which would eliminate entirely the need for the 3,000 staffers at the PBS division.

And surely next in line would be the 2,000 staff in the "Staff Offices" division. This operation encompasses a mini-overhead department with policy and planning functions, legal and regulatory affairs, human resources, financial management and budgeting, communications and public affairs, information technology, facilities management, acquisition support, and emergency preparedness sub-divisions and offices.

For crying out loud, this amounts to a bureaucratic version of the Russian Matryoshka Dolls–the same function stacked inside the next and the next and the next. All of the client agencies of both the GSA and OPM have these same functions. And if these "Staff Offices" functions don't primarily service client agencies, which have their own overhead functions, the story is even worse: The 2,000 bureaucrats in these overhead activities amount to one for every *6.5* GSA employees.

Distribution of GSA 13,000 Staff

Function/Activity	Estimated Staff Count
Federal Acquisition Service (FAS)	5,000
Public Buildings Service (PBS)	3,000
Staff Offices	2,000
Regional Offices	3,000

Indeed, even when it comes to the 5,000 staffers at the Federal Acquisition Service (FAS), it is evident that we have another case of centralized support functions run amok within the vast labyrinth of the Federal government. Among the nine major functions of the FAS division, upwards of eight of them surely duplicate thousands of bureaucrats performing the same functions in the client agencies. The only function logically needed at the GSA is the policy and compliance function, which, again, could be done by a few dozen overseers.

The argument could be made, of course, that the duplication should be eliminated at the client department and agency level rather than GSA. But as the man said, the best way to kill a snake is to smash the eggs.

- **Technology Transformation Services (TTS):** Leads the digital transformation of the federal government by helping agencies build, buy, and share technology that allows them to provide more accessible, efficient, and effective products and services for the American people.
- **Office of Assisted Acquisition Services (AAS):** Provides acquisition, technical, and project management services that assist agencies in *acquiring and deploying information technology* and professional services solutions.
- **Office of General Supplies and Services (GS&S):** Simplifies the buying process to meet customer agency needs in the areas of *office supplies, computer products, tools, security, furniture*, and many others.
- **Office of Information Technology Category (ITC):** Offers high-quality commercial products, systems, services, and support to federal, state, and local governments.
- **Office of Professional Services and Human Capital Categories (PSHC):** Responsible for the strategic leadership, oversight, and management of FAS' professional services, human capital services, and ***charge card*** management programs and contracts.
- **Office of Travel, Transportation, and Logistics Categories (TTL):** Develops and manages programs and solutions for ***travel, motor vehicle, and transportation services,*** and FAS emergency support.
- **Office of Policy and Compliance:** Ensures operating practices are consistent and that GSA's activities are fully

compliant with federal laws, regulations, and policies.

- **Office of Customer and Stakeholder Engagement:** Enables the GSA to better understand customer requirements and become a strategic partner in helping agencies meet their acquisition needs.
- **Office of Strategy and Innovation:** Sets the strategic vision for all of FAS, consistent with the overall GSA strategic vision, and also a consistent set of tools to help achieve the vision.

In short, a $1 billion savings from eliminating 6,400 staff positions (50%) would be a minimal start on reducing the fat at GSA. But billions more could be eventually saved by selling-off redundant GSA-owned buildings and subletting what would be millions of unused leased square footage to private sector users.

CHAPTER 4

Pro Rata Staff Cuts of 34% At All Other Nondefense Agencies

Anyone confused or delusional about the difficulty of making even modest headway against the nation's towering public debt–to say nothing of the worthy and necessary DOGE goal of $2 trillion per year in budget savings–should contemplate the following. To wit, the "Slash the Fat" portion of our DOGE plan amounts to just $400 billion or only 20% of the savings target. Yet after reviewing the shutdown of 16 agencies including biggies like the FBI, OSHA, AID, and the Education Department, plus another 9 agencies targeted for staff cuts of 50% including the IRS, EPA, and NASA, the combined savings are modest indeed.

The 164,000 headcount reductions and $26 billion of fully-loaded compensation savings from cutting staff waste and excess at these 25 agencies would amount to just–

- *6.5%* of the $400 billion "Slash the Fat" target.
- *1.3%* of the $2 trillion overall DOGE target.

- *0.33%* of the $8.0 trillion of total Federal spending projected by CBO for the FY 2029 target year.

The idea of "more wood to chop," therefore, hardly does justice to the challenge at hand. Still, as we indicated above, a ***40% cut*** in the headcount of all nondefense Federal agencies, or about 535,000 staff positions, is a more than reasonable goal. That is, given the threat of runaway public debt, as well as the precedent at X where Elon Musk was able to shrink staffing by 80% without missing a beat in terms of company operations and customer service.

Accordingly, an additional ***371,000 job cuts*** would be required in the balance of the nondefense departments and agencies to achieve the 40% staff reduction. In turn, such reductions would generate about ***$59 billion*** in fully-loaded employee compensation cost savings by the FY 2029 target year.

To be sure, given the inherent inefficiency and payroll padding, which is endemic in the Federal bureaucracy, this additional cut of 371,000 positions would represent just 34% of the ***1.087 million*** nondefense employees outside of the 25 agencies itemized previously for deeper reductions. So it is more than achievable, but it would also generate thundering political opposition–especially given the motherhood character of the main agencies where the cuts would have to be taken.

Nevertheless, the necessary headcount and compensation cost savings for the target year of FY 2029 would then stack up as follows for the nondefense side of the Federal government overall:

- 16 Agencies Eliminated: 71,000 staff cuts and $11 billion of savings.

- 9 Agencies with 50% Staff Reductions: 93,000 staff cuts and $15 billion savings.
- 34% Pro Rata Cut In all Other Agencies: 371,000 staff cuts and $59 billion savings.
- Total, Nondefense Staff Reductions: 535,000 staff cuts and $85 billion savings.

With respect to the balance of the Federal government and an average staff cut of 34%, the bulk of the reductions would need to come from the seven departments listed below. These units account for the overwhelming share of nondefense employment outside of the 25 agencies slated for deeper staff reductions:

Current Headcounts at Major Nondefense Agencies:

- Veterans Administration: 436,500.
- Homeland Security: 240,000.
- Agriculture Department: 100,000.
- HHS Department: 80,000.
- Interior Department: 70,000.
- DOJ ex-FBI, DEA, and ATF: 65,000.
- Social Security Administration: 60,000.
- Total, 7 Largest Agencies: ***1,051,500.***

Needless to say, cutting the headcount at the largest of these, the Veterans Administration, by the indicated 34% factor, should be a cakewalk. That's owing to the inherent massive inefficiency of the Veterans health care and hospital system. The latter makes even the inefficiency-ridden US civilian hospital system seem like a paragon of efficiency by comparison.

Unfortunately, none of the powerful veterans organizations or their dutiful water carriers on the House and Senate Veterans committees see it that way. For instance, any attempt to substantially outsource the work of the VA's 170 Medical Centers (hospitals) or 1,200 outpatient clinics and treatment facilities to more efficient private-sector providers would generate a firestorm of political opposition. Yet these two divisions alone account for 80% of the VA's ***436,500 headcount***.

Still, the implied inefficiency of the VA hospital system (VAMC) borders on the astounding. There are currently 70,000 hospital beds in the VA Medical Centers network, which tend to be 65% to 70% occupied on an average basis. This means that the staff ratio for VMAC's 295,000 employees is ***4.4 staff*** for every bed and ***6.0 staff*** for every patient at the average occupied census. Were VAMC's 295,000 employees to file into the VA's hospital rooms with 49,000 beds occupied at any moment, therefore, there would be standing room only!

Moreover, the other large piece of the VA headcount consists of the 31,000 jobs in the Benefits Administration. The latter paid out $192 billion in veterans disability compensation and pensions in FY 2023 to 5.6 million veterans–96% of whom are disabled. The issue here is that disability payment determination and processing is labor-intensive because payment levels are based on a complex system of disability ratings.

Still, the current fully-loaded compensation cost for the Benefits Administration staff at $140,000 per head is $4.34 billion, or ***2.3%*** of benefit payments disbursed. By contrast, the Social Security Administration (SSA) paid out more than six times more benefits in 2023 ($1.190 trillion) with only double the staff (60,000). So the SSA staff overhead ratio was just ***0.7%***.

At the end of the day, of course, there is no practical alternative to a fully warranted, sweeping headcount reduction at the VA. That's because at the 34% target for all other nondefense agencies, the required VA staff reduction would total ***148,000 employees***, meaning that the VA contribution alone would also comprise ***40%*** of the 371,000 target for all other nondefense agencies.

That is to say, if VA is let off the hook by its powerful constituencies, it is simply not realistic to think that overall nondefense employment can be cut by 535,000 jobs and $85 billion of compensation costs. Indeed, if a big agency like the VA gets a hall pass, the whole effort at massive Federal staff reductions would quickly unravel.

Breakdown Of VA Headcount By Major Function

Division	Estimated FTE Count
Veterans' Health Administration (VHA)	403,000
-- VA Medical Centers (VAMCs)	295,000
-- Community-Based Outpatient Clinics (CBOCs)	55,000
-- Other Healthcare Units	53,000
Veterans Benefits Administration (VBA)	31,000
National Cemetery Administration (NCA)	2,000
Other VA Offices and Support Functions	500
Total	**436,500**

That would be true in spades if the next giant nondefense employer–the Homeland Security Department with ***240,000*** employees–were also to be let off the hook. Yet, as shown below, 80,000 of the agency's 240,000 headcount is in the Customs and Border Protection (CBP) and ICE–both of which the Trump Administration intends to increase sharply.

Yet this would self-evidently create an insuperable math problem. If CBP and ICE staffs were to be increased by the rumored 20% (+16,000 jobs) while the overall Homeland Security Department was cut by the pro rata target of 34% (-81,000 jobs), there would be room for only 63,000 Homeland headcounts outside of CBP and ICE.

At present, however, the Homeland Security headcount outside of CBP/ICE totals 160,000, including 50,000 in the TSA, 40,000 in the Coast Guard, 15,000 in FEMA, and 7,000 in the Secret Service. To shrink the Homeland Security Department headcount by the 34% pro rata target, therefore, would require a 97,000 or 61% headcount reduction in these and the other smaller Homeland Units.

Notwithstanding these obstacles, however, major headcount reductions at the Homeland Security Department would be readily possible. For instance, a 25,000 chunk of the 81,000 of needed headcount reductions at the Homeland Security Department could be achieved at the TSA if they used modern technology rather than requiring people to take off their shoes, hats, and belts at airport security gates.

Likewise, a 50% cut at FEMA and the Coast Guard would provide another 23,000 of staff reductions--notwithstanding the screaming political opposition that would come out of the woodwork at the relevant Congressional committee hearings. And surely the balance of 48,000 current staff positions in the overhead areas shown in the last four lines below could contribute another 33,000 of the needed reduction.

In all, the three above-mentioned areas would generate the needed 81,000 Homeland staff cuts. Of course, that would leave no room at all for the proposed CBP and ICE staff increases, but as we have explained elsewhere these increases are not really necessary. The correct answer is to raise not the staff count but the quota for legal immigration of unskilled workers from a pittance of 4,000 per year to say, 500,000 annually. That would cause 90% of the congestion at the border to disappear because applicants could get visas at local embassies and State Department consulates at dozens of locations in Latin America rather than coming to the US border to break the law and get in the queue for asylum.

In short, you wouldn't even need 20,000 more employees at CBP and ICE if you had a rational-legal program to vet and process badly needed low-skill workers through the already existing legal immigration machinery of the State Department. The problem is the rigid immigration quota system in the context of a drastically labor-short US economy, not an insufficient number of gendarmes at the Rio Grande.

Breakdown Of Homeland Security Department's FTE Count

Agency	Estimated FTE Count
U.S. Customs and Border Protection (CBP)	60,000
U.S. Immigration and Customs Enforcement (ICE)	20,000
Transportation Security Administration (TSA)	50,000
Federal Emergency Management Agency (FEMA)	15,000
U.S. Coast Guard (USCG)	40,000
U.S. Secret Service (USSS)	7,000
Cybersecurity and Infrastructure Security Agency (CISA)	2,500
Office of Intelligence and Analysis (I&A)	1,000
Office of Operations Coordination (OPS)	1,500
Other DHS Offices and Support Functions	43,000
Total	**240,000**

Another important nut to crack in the 34% staff reduction category would be the Agriculture Department (USDA). Surely, all the little fiefdoms shown below could be cut by 34% with little harm to the public good. Indeed, functions like taxpayer-funded agricultural research, subsidized credit via the Farm Service Agency, the overseas marketing forays of the Foreign Agricultural Service, and the domestic market manipulations of the Agricultural Marketing Service – should all be abolished entirely. Those functions are relics of the New Deal and even earlier but have never been the proper function of the state.

Still, implementation of these well-warranted policy measures would reduce the USDA headcount by just 18,000. Accordingly, another 16,000 would need to come out of the Forest Service and the USDA overhead departments if food safety and APHIS are to be manned at near current levels.

Again, from a policy perspective, 34,000 jobs can be easily cut from the politics-ridden USDA. But the department is controlled lock-stock-and-barrel by the farm lobbies via their handmaids on the Congressional Ag committees.

So getting the savings would require a concerted application of political resolve from the White House, which history has shown to be achievable. Thus, when Ronald Reagan came to the White House in January 1981 the headcount at the USDA was 144,000, but by 1983 we had it down to 124,000 and then fell further to 113,000 by 1986. That amounted to a **27% staffing reduction**–or nearly what would be needed again in the current instance.

Breakdown of FTEs at USDA for FY 2023

Division	Estimated FTE Count
Agricultural Research Service (ARS)	8,000
Animal and Plant Health Inspection Service (APHIS)	7,000
Food Safety and Inspection Service (FSIS)	9,000
Forest Service (FS)	35,000
National Institute of Food and Agriculture (NIFA)	1,000
Economic Research Service (ERS)	1,000
National Agricultural Statistics Service (NASS)	1,000
Farm Service Agency (FSA)	5,000
Natural Resources Conservation Service (NRCS)	4,000
Risk Management Agency (RMA)	1,000
Rural Development (RD)	4,000
Foreign Agricultural Service (FAS)	2,000
Agricultural Marketing Service (AMS)	3,000
Office of the Secretary (OST)	2,500
Office of Inspector General (OIG)	1,500
Other USDA Offices and Support Functions	15,000
Total	**100,000**

The Department of Health and Human Services sports a current payroll of 80,000–so the 34% pro rata cut would require a 27,200 reduction in headcounts. Again, there is plenty of overhead and small bureaucratic duchies that could be cut by one-third–most especially the 6,000 headcounts in the Office of the Secretary and related support functions.

However, the preponderant share of the HHS payroll is accounted for by the CDC (10,000), FDA (18,000), and the NIH (20,000). Given the lead role these three agencies played in foisting the destructive Covid panic, lockdowns, and mandatory vax shots on the nation, their total headcount of 48,000 should be reduced by one-third simply as a matter of reparations for the vast harm these agencies have done since February 2020.

In all, there is plenty of room for a 27,200-headcount reduction at HHS. It would save upwards of $3.7 billion by the target year of FY 2029. And that's before the related overhead and program savings outlined in another section below.

Department of Health and Human Services (HHS) Headcounts *for FY 2023*

Division	Estimated FTE Count
Administration for Children and Families (ACF)	8,000
Administration for Community Living (ACL)	2,000
Agency for Healthcare Research and Quality (AHRQ)	1,000
Agency for Toxic Substances and Disease Registry (ATSDR)	1,000
Centers for Disease Control and Prevention (CDC)	10,000
Centers for Medicare & Medicaid Services (CMS)	6,000
Food and Drug Administration (FDA)	18,000
Health Resources and Services Administration (HRSA)	3,000
Indian Health Service (IHS)	2,000
National Institutes of Health (NIH)	20,000
Substance Abuse and Mental Health Services Administration (SAMHSA)	2,000
Office of the Secretary (OS)	5,000
Office of Inspector General (OIG)	1,000
Other HHS Offices and Support Functions	1,000
Total	**80,000**

Another tough row to hoe would be in the residual of the Department of Justice after setting aside the 50,000 jobs in the FBI, DEA, and ATF which were treated separately on the list of 16 agencies to be abolished. The remaining payroll of 65,000 sounds large, but fully **40,000** of that is accounted for by the Bureau of Prisons, along with 4,000 in the US Marshals Service, 5,000 in the US Attorneys division, and 1,200 in the Criminal Division.

Accordingly, these three law enforcement-focused divisions account for nearly 80% of the 65,000 headcounts in the DOJ outside of the three agencies to be abolished. Finding a 34% pro rate cut or 22,000 reductions would be a challenge, but the place to start would obviously be the Federal prison system.

At the present time the inmate population of the Federal Bureau of Prisons (BOP) is 157,806, but 14,030 of these are held in contract prisons run by private operators. So the net population actually incarcerated in the 111 BOP prisons is 143,377, meaning that ***BOP has one employee for every 3.6 inmates!***

Moreover, about 64,500 of the current BOP inmates are incarcerated on drug charges of all types, and upwards of half of those are for marijuana and small use or distribution offenses related to other illicit drugs. So if half of the BOP prison population serving under drug possession and other social "crimes" were released for time served and the BOP employee-to-inmate ratio was raised to 6:1, ***the BOP headcount could be cut in half!***

In any event, even the rump of the DOJ--which is due for a sweeping housecleaning–could find the ways and means to make its 34% pro rata cut, thereby contributing a 22,000-headcount reduction to the effort.

In the case of the 60,000 employees of the Social Security Administration, the breakdown below suggests that the agency is top-heavy in the extreme.

Social Security Administration (SSA) Headcounts *for FY 2023 by function (total: 70,000):*

Division	Estimated FTE Count
Office of Central Operations	10,000
Office of Central Operations Support	5,000
Office of Disability Adjudication and Review (ODAR)	15,000
Office of Hearings Operations (OHO)	5,000
Office of Retirement and Disability Policy (ORDP)	5,000
Office of Inspector General (OIG)	2,000
Field Offices	20,000
Program Service Centers	2,000
Regional Offices	1,000
Other SSA Offices and Support Functions	5,000
Total	

Only 20,000 employees are in the field offices and another 15,000 are in the Office of Disability Adjudication and Review and 5,000 are in Hearings Operations (OHO). These recipient-facing staff could certainly be organized to function in a more efficient manner, especially if provided with more effective computer-based assistance.

At the same time, the other 20,000 employees consist mainly of overhead and support functions. A third of these could likely be eliminated with no diminution of service to the SSA's 70 million beneficiaries.

Finally, the payroll at the Interior Department totals *70,000.* The 34% pro rata reduction would total ***23,800*** and generate a total compensation savings of $3.2 billion per year. Based on the breakdown of current Interior Department headcounts, achieving a reduction of that magnitude would appear to be eminently doable.

For instance, the 10,000 headcount in the office of the Secretary and related overheads amount to one top-side bureaucrat for every 6.0 employees in the operating bureaus.

Likewise, there are 6,000 staff in the Indian Affairs and Indian Education bureaus or one bureaucrat for every 400 Indians served by the Interior Department*!*

Surely that ratio could be increased materially, assuming the paternalistic BIA and BIE do anything at all to actually better the lives of their clients.

Full-time equivalent (FTE) employees at the U.S. Department of the Interior (DOI) for FY 2023:

Division	Estimated FTE Count
Bureau of Land Management (BLM)	10,000
Bureau of Ocean Energy Management (BOEM)	1,000
Bureau of Safety and Environmental Enforcement (BSEE)	1,000
Office of Surface Mining Reclamation and Enforcement (OSMRE)	1,000
Bureau of Reclamation (BOR)	5,000
U.S. Geological Survey (USGS)	10,000
Fish and Wildlife Service (FWS)	10,000
National Park Service (NPS)	15,000
Bureau of Indian Affairs (BIA)	5,000
Bureau of Indian Education (BIE)	1,000
Office of the Secretary (OS)	5,000
Office of the Solicitor (SOL)	1,000
Office of Inspector General (OIG)	1,000
Other DOI Offices and Support Functions	4,000
Total	**70,000**

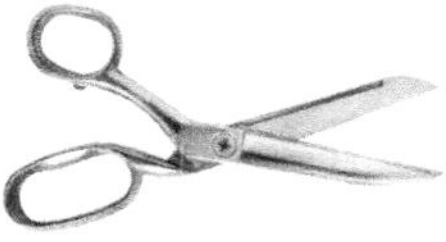

CHAPTER 5

Slash The Fat: $45 Billion Of Overhead Savings

In the exercise of cutting 40% of the nondefense payroll, of course, we slew a veritable herd of sacred cows. This encompassed the shutdown of 16 agencies including biggies like the FBI, OSHA, AID, and the Education Department. We also targeted another nine heavily bloated bureaucracies for staff cuts of 50% including the IRS, EPA, and NASA.

Finally, we laid out the impact and savings from a 34% pro-rata staffing cut (371,000) in all the other Federal nondefense agencies. Among others, the latter included a 148,000 headcount reduction at the Veterans Administration, 81,000 in the Homeland Security Department, and 34,000 at the Department of Agriculture.

But here's the thing. We are just getting started!

The entirety of these 535,000 staff reductions would save just ***$85 billion***. And that's based on a generous assumption that by the target year of FY 2029 the average fully-loaded cost of a nondefense FTE (full time equivalent) would be $160,000 per year including pay, fringes, health care, pensions, and other benefits.

Still, the total budget savings as tallied below would amount to just 21% of our "Slash the Fat" target of $400 billion and just 4% of the overall DOGE goal of $2 trillion in annual budget savings.

FY 2029 Nondefense Headcount Cut and Savings Summary:

- 16 Agencies Eliminated: 71,000 staff cuts and $11 billion of savings.
- 9 Agencies with 50% Staff Reductions: 93,000 staff cuts and $15 billion savings.
- 34% Pro-Rata Cut in all Other Agencies: 371,000 staff cuts and $59 billion savings.
- Total, Nondefense Staff Reductions: ***535,000 staff cuts and $85 billion savings.***

Fortunately, however, eliminating entire agencies and their missions or deep staff reductions and curtailment of related activities at all other government agencies would generate substantial collateral non-compensation savings. This includes both reduced internal operating items–like occupancy, heat, light, supplies, travel, training, etc.—-but also curtailment of outside spending for contractors, promotional/propaganda campaigns, and most especially grants to client organizations and state and local governments.

For instance, the fully-loaded staff savings from shutting down the Legal Services Corporation in FY 2023 would have been $112 million, but its total budget that year was $675 million. The balance went to non-compensation operating costs and grants to NGOs and other legal aid entities. So eliminating the balance of non-compensation spending and inflating it to an FY 2029 level would save an additional ***$530 million*** per year at the Legal Services Corporation alone.

Likewise, the $42 million of combined staff compensation savings from eliminating the National Endowments for the Arts and Humanities and the Corporation for Public Broadcasting would generate an additional ***$975 million*** of savings owing to the elimination of operating overheads and the extensive grant-making activities of these agencies.

In the case of terminating what we have described as the "Empire First" international operations–the Agency for Global Media and the National Endowment for Democracy–the combined 2029 staff compensation savings of $206 million would be accompanied by a further ***$1.1 billion*** per year of non-payroll savings. This would result from eliminating extensive operating costs across the globe, as well as hundreds of millions worth of walking around money handed out each year by these agencies to clients all over the planet in pursuit of Washington's hegemonic policies.

In short, our "Slash the Fat" plan for these two agencies alone would save a total of $1.31 billion by FY 2029. And these outlays truly are "fat, waste, and abuse" because spreading the gospel of democracy, long pants, and Coca-Cola to the far corners of the planet is not the proper business of the state at all–to say nothing of its sheer affront to the hard-pressed taxpayers of Flyover America who are forced to foot the bill.

Likewise, the elimination of NHTSA and its 600 staff would save just $96 million in fully-loaded compensation costs, but would also eliminate the agency's huge overhead–along with upwards of $500 million per year of safety grants to the states help them inform motorists, for example, not to text while they are driving!

The collateral savings from terminating the agency and all of its unnecessary and inappropriate Federal functions, therefore, would total ***$1.4 billion*** per year by FY 2029.

However, the big non-payroll savings would come from getting the Federal government out of the law enforcement, drug persecution, and gun control businesses. These are police functions that the Founders intended to be carried out by the states and localities, as the 10th Amendment made perfectly clear:

> *"The powers **not delegated** to the United States by the Constitution, nor prohibited by it to the States, are reserved to the States respectively, or to the people."*

Yet due to the massive operational expense of the FBI, DEA, and BATF, the ***$6.8 billion*** compensation cost savings from eliminating these three Federal agencies would be just the down payment. When you add in their massive networks of facilities, vehicle fleets, weaponry, outside contractors, and grants to nonprofits and other local police agencies, the additional savings would amount to ***$8.3 billion.***

Overall, in the case of the 16 terminated agencies, the collateral savings would total $17 billion by 2029 or ***150%*** of the $11 billion in staff compensation reductions.

Additional Non-Compensation Savings by FY 2029 for 16 Terminated Agencies:

- FBI: $5,500 million.
- DEA: $2,100 million.
- BATF: $760 million.
- NHTSA: $1,400 million.
- Agency for Global Media: $750 million.
- Legal Services Corporation: $650 million.
- Education Department: $1,050 million.
- AID: $2,450 million.
- Other Shutdown Agencies: $2,493 million.
- Total, 16 Terminated Agencies: ***$17,000 million***

In the case of the 9 agencies with 50% staff reductions, the collateral savings ratio would be considerably lower because much of the operating cost, infrastructure and outside contracting, and spending activities would be retained, albeit in sharply thrifted form. Accordingly, we estimate that additional savings at these nine agencies including the IRS, EPA, and NASA would average ***55%*** of the compensation cost savings.

Again, the largest share of the $8 billion of non-payroll savings at these nine agencies would be accounted for by the IRS ($3.65 billion), the FAA ($1.98 billion), and NASA ($800 million). In every sense of the word, these massive expenditures for occupancy, supplies, travel, outside contractors, and promotion and propaganda constitute "fat, waste, and excess" because they serve no legitimate interest of the state.

Additional Non-Compensation Savings by FY 2029 For 50% Cut Agencies:

- FAA: $1.98 billion.
- IRS: $3.65 billion.
- NASA: $800 million.
- Six other 50% cut agencies: $1.57 billion.
- Total, 50% cut agencies: ***$8.0 billion.***

As indicated above, the dozens of remaining nondefense departments and agencies would be assigned a pro-rata headcount cut of 34%. Here, too, the collateral savings due to the elimination of 371,000 bureaucrats would be considerable. But these savings in agency overhead and outside contracts and other spending would accrue at a lower fraction of compensation cost savings owing to the fact that 70% of the activities of these agencies would continue.

Accordingly, we cautiously assume a ***33%*** ratio to the $59 billion of payroll savings in this sector, which translates into ***$20 billion*** of non-payroll cost savings.

Again, $12 billion of this total would be attributable to the big three nondefense employers–the Veterans Department, Homeland Security Department, and the USDA. So if the powerful constituencies and lobbies that protect these sacred cow departments succeed in blocking their targeted 263,000 staff cuts, the projected non-payroll budget savings would deflate rapidly.

Additional Non-Compensation Savings by FY 2029 For 29% Staff Cut Agencies:

- Veterans Department: $6.7 billion.
- Homeland Security Department: $3.7 billion.
- Agriculture Department: $1.53 billion.
- HHS: $1.2 billion.
- Interior Department: $1.07 billion.
- Social Security Administration: $918 million.
- DOJ ex-FBI, DEA and BATF: $1.0 billion.
- Other 29% cut agencies: $3.88 billion.
- Total, 29% staff cut agencies: ***$20.0 billion.***

In short, all of these savings are fully justified and eminently doable as a practical matter, but the legislative and political hurdles cannot be gainsaid. Even then, however, the combined payroll and non-compensation costs savings from our "Shrink the Fat" exercise would amount to just ***$130 billion by 2029.*** And that would leave another $270 billion to be found in the fraud, waste, and abuse embedded in other areas of the budget and tax code, as we will outline below.

Summary of "Slash the Fat" Savings From Cutting 40% of the Nondefense Payroll in FY 2029:

- Compensation Savings from 16 Terminated Agencies: $11 billion.
- Compensation Savings From 9 Agencies Cut by 50%: $15 billion.
- Compensation Savings from Cutting the Balance of

Nondefense Agencies by 34%: $59 billion.

- Non-compensation Savings From 16 Terminated Agencies: $17 billion.
- Non-compensation savings from 9 agencies cut by 50%: $8 billion.
- Non-compensation Savings from All Other Nondefense Agencies: $20 billion.
- Total, Compensation and Collateral Savings: ***$130 billion.***

CHAPTER 6

$270 Billion Savings From Special Interest Fat, Waste and Abuse

Pro-government statists have used a tried-and-true playbook for decades. They simply claim that certain sectors of the private economy suffer from malfunctions, imperfections, and other alleged market failures, and then insist that the state intervene with compensatory fiscal subsidies and/or regulatory assists.

Perhaps the most egregious example of this well-honed swindle is the ***$2.5 billion*** per year currently being spent on EV (Electric Vehicle) credits of $7,500 per qualifying vehicle. Needless to say, the government has absolutely no business attempting to tilt the mix of EV versus IC (internal combustion) vehicles sold in the marketplace. And borrowing money to fund these credits, as, on the margin, Washington currently does, is especially egregious because fully 83% or $2 billion of this taxpayer subsidy goes to households with more than $100,000 of income.

Of course, these EV credits are just the tip of the iceberg.

So we estimate that the balance of the $270 billion needed to achieve our "Slash the Fat" target of $400 billion per year in budget savings could easily be achieved by repealing a wide range of the like and similar "fraud, waste, and abuse" subventions that Washington has showered upon the beltway lobbies and squeaky wheels of the private sector. In particular, there are three sectors that account for most of this additional form of fat and waste:

- agriculture.
- energy.
- other business special interest lobbies.

In all these cases, the target industries and private sector beneficiaries have been falsely depicted as especially virtuous or crucial to national economic function and well-being. In turn, this supposedly makes them worthy of special dispensations from the fiscal coffers and regulatory powers of the state–a proposition we heartily dispute.

The $40 Billion per Year Farmer Subsidy Waste

Perhaps the foremost case of this false valorization and subsidization is that of farmers. They are depicted as solidly virtuous yeoman who must contend with the slings and arrows of Mother Nature and the travails of long hours of work under the blazing sun–even as their economics are purportedly jeopardized by volatile, unpredictable commodity prices and the episodic threat of crop failure owing to weather and pestilence.

Well, no, farmers are neither virtuous nor special. They are just regular folks who happen to have an outdoors job and who face economic challenges and risks that are no different than those confronting hundreds of other occupations and industries. We are referring, for instance, to the adversities faced by snowplow and ski-slope operators in a no-snow year, umbrella manufacturers during a drought, and resort operators during a recession, among countless others.

Stated differently, the principles of free market capitalism are applicable always and everywhere but especially in agriculture. At a price, farmers can insure against the weather–it's part of the cost of the corn, wheat, and cattle business. They can also base their economics on averaging the cycle of high, low, and middling prices that have recurred in the commodity markets from time immemorial–just as do unsubsidized producers in energy, mining, timber, and other resource extraction industries.

So there is zero case for current farm price supports, subsidized crop insurance, conservation payments or indirect subsidies for agricultural research, and farm product marketing and exports. Indeed, when you compare the average incomes of farm households (both on-farm and off-farm earnings) with the national average household income, the case is crystal clear. In the last year for which comparable data were available (2017), the average farm household income of $113,500 was about *27% higher* than the average U.S. household income of $89,000.

Moreover, the gap has actually been widening over recent decades. Since 1960, inflation-adjusted farm incomes (black line) are up by ***350%***. That's far greater than the average ***120%*** gain in real terms for all US household incomes (red line).

Figure 28. Farm Household Income Has Been Above U.S. Average, 1960-2018

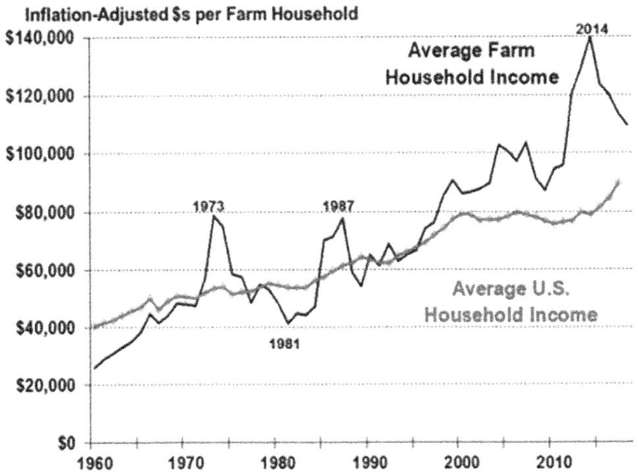

Overall, these pro-farmer forms of fraud, waste, and abuse total about $40 billion per year, as detailed below. Yet they persist owing to a log-rolling coalition between farm interests and urban and Dem-based pro-welfare advocates, whose coalition has completely hijacked the USDA budget. In effect, the $133 billion per year worth of welfare programs (Food Stamps, school lunches, and WIC) currently run out of the USDA is the dog that wags the $40 billion farmer subsidy tail within the department's massive $189 billion budget (FY 2023).

Breakout of 2023 USDA Budget:

- Food Stamps: $112.8 billion.
- School Lunch Program: $14.0 billion.

- Women, Infants and Children (WIC): $6.5 billion.
- ***Subtotal, Welfare programs: $133.3 billion.***
- ***Farmer Subsidies: $40.0 billion.***
- Food Safety and other USDA functions: $15.7 billion.
- USDA Budget Total: $189.0 billion.

None of the farmer subsidies can be justified as a matter of economics or the proper role of government. As shown below, the $40 billion goes mostly to production and conservation subsidies which are relics of the New Deal and are utterly unnecessary for the functioning of the corn, wheat, soybean, cotton, dairy, and peanut industries which get the loot.

For instance, during FY 2023 corn farmers got $2.5 billion from Uncle Sam to prop up prices, thereby accounting for just 3.4% of the value of the $73 billion US corn crop that year. Similarly, grower subsidies amounted to 9% of the $13 billion wheat crop, 11% of the $5 billion cotton crop, and lower percentages for the others. In short, farm subsidies are just marginal frosting on the cake–modern farming could function with market-based income alone.

Moreover, the total value of farm production in 2023 was just $204 billion or only a tiny *0.7%* of GDP. Accordingly, there is no possible way that any marginal reduction in the value of farm output and income owing to inviting farmers to rejoin the ranks of free market producers would have any noticeable impact on the nation's economic welfare. Perhaps what it might do, instead, is raise the question as to why 0.7% of GDP gets its very own US government Cabinet Department, 100,000 bureaucrats, and Ag committees in both houses which never stop confecting new ways to fleece the taxpayers.

FY 2023 Price Support Subsidies

Commodity	Budget (in billions)
Corn	$2.5
Wheat	$1.2
Soybeans	$1.0
Cotton	$0.5
Dairy	$0.8
Peanuts	$0.2
Sugar	$0.3
Other Commodities	$2.7
Total	**$9.2**

Likewise, there is no more reason for the state to promote and subsidize farm exports than there is for promoting the export of washing machines, recycled cardboard, or polyethylene. All domestic producers are entitled to earn what they can on the global markets, but it is a pure abuse to charge the taxpayers for the kinds of marketing expenses that Whirlpool, America Chung Nam, or Exxon pay out of their own pockets, respectively, to promote the aforementioned products abroad.

The same is true of the $10 billion that goes to rural development and agricultural research, education, etc. There is no better reason for these "rural" programs than for similar research and education programs focused on urban, suburban, exurban, shoreline, desert, or mountain range communities.

The dollars spent for the purposes summarized below are owing to *political muscle* in the halls of Congress, and nothing more. They are just as much an example of waste and abuse as the proverbial exposes of wasteful research grants, such as the

$3 million to study why hamsters fight or the $518,000 study on how cocaine affects the sexual behavior of Japanese quails.

Distribution of Farmer Subsidies in 2023

- Production and conservation subsidies: $28.0 billion.
- Trade and Foreign Marketing subsidies: $2.0 billion.
- Rural developments subsidies: $4.0 billion.
- Agricultural research, education and related: $6.0 billion.
- ***Total Farm subsidies: $40 billion.***

Within the $28 billion of subsidies for production and conservation, the waste and abuse is starkly transparent. In addition to the aforementioned $9.2 billion of commodity price supports, for instance, a total of $7 billion was spent in 2023 for crop insurance subsidies and other risk mitigation handouts to the major commodity growers, as well as to specialty producers of fruits, nuts, and vegetables. The only "farm" crop that doesn't get these handouts, apparently, is the $5 billion marijuana industry, which seems to do just fine without Uncle Sam's helping hand and after being "debanked" by the Feds, to boot.

Similarly, the $4 billion spent by the Farm Service Agency goes mainly to deeply subsidize interest rates on loans to farmers. These below-market government loans and guarantees can be used to purchase or expand a farm or ranch, pay closing costs, construct or improve buildings, and help conserve and protect soil and water resources. From another loan basket, farmers can also get subsidized "operating loans" for the purchase of livestock, seed, fertilizer, equipment, etc. They can also cover family living expenses while a farm is getting started!

Well, for crying out loud. Why do farmers get a 4.5% rate on operating loans courtesy of the taxpayers when the current prime rate according to Bankrate is 8.5% for all other businesses in the private sector? The fact is, tree-trimming services and landscaping operators work with their hands out in the elements under uncertain economic conditions, too. But with no help from the taxpayers.

However, the latter's failure to wrangle a Washington-bestowed exemption from the rigors of the free market isn't owing to lesser virtue. Alas, the difference with farmers is purely a consequence of political muscle on the banks of the Potomac.

The $8 billion for a half-dozen varieties of "conservation" programs beg the same question. That is to say, if various farmland conservation practices are economic, they wouldn't need subsidies from Washington. And if they are mainly designed to provide a nicer habitat for quail and jackrabbits, why should the taxpayers of Maine care about alleged soil erosion and/or farmer encroachment on wildlife habitats in Iowa?

Let local taxpayers fund such public amenities, if they are to be funded at all.

USDA Production and Conservation Subsidies

Program	Budget (in billions)
Farm Service Agency (FSA)	$4.0
Risk Management Agency (RMA)	$2.0
Natural Resources Conservation Service (NRCS)	$3.0
Conservation Reserve Program (CRP)	$1.8
Agricultural Conservation Easement Program (ACEP)	$0.5
Environmental Quality Incentives Program (EQIP)	$1.5
Regional Conservation Partnership Program (RCPP)	$0.3
Conservation Stewardship Program (CSP)	$0.7
Commodity Credit Corporation (CCC)	$9.2
Federal Crop Insurance Program	$5.0
Other Conservation Programs	$0.6
Miscellaneous and Other Expenses	$0.4
Total	**$28.0**

At the end of the day, getting rid of the $40 billion farmer subsidies is a no-brainer, but also something else. To wit, it's another litmus test as to whether the DOGE goal of $2 trillion of annual budget savings has even a snowball's chance in the hot place. After all, eliminating $40 billion of farmer subsidies that are grounded in 80 years of politics on the banks of the Potomac would only take our cumulative savings of $130 billion, due to pink-slipping 535,000 of government bureaucrats, to $170 billion. And that's still only 8.5% of the way to the $2 trillion DOGE goal.

In this context, there is one surefire way to facilitate the elimination of this $40 billion of wasteful farmer subsidies. ***The Trump White House should firmly, unequivocally, and militantly announce that it will not sign any log-rolling farm bill in any way, shape, or form, and that the welfare/***

food programs must be separated legislatively from the farm subsidy programs up for reauthorization next year.

Only that kind of decisive break from past practice can enable the legislative process to proceed in an honest manner, untainted by corrupt political deals.

The Energy Boondoggles

Back in the day as a member of the Energy Committee in the US House and at OMB we fought the 1970s energy subsidy boondoggles tooth and nail because they were anti-free market but also based on a sheer falsehood. That is, the notion that the world was running out of fossil fuels rapidly and that only a state-orchestrated transition to renewables and artificial conservation intermediated by heavy subsidies, tax credits, and regulatory mandates could save the nation from ruin.

As it happened, in the early 1980's we succeeded in eliminating most of these Nixon-Ford-Carter energy market interventions. But these actions did not lead to ruin at all. That's because fossil fuel depletion was a complete myth, which myth failed to grasp that higher prices and better technology would substantially expand the "economically recoverable" fossil resource base. As it happened, therefore, global oil and natural gas production rose from 85 million BOEs/day (barrels of oil equivalent) in 1980 to 130 million BOEs/day in 2023.

This *53%* increase in production, of course, did not stop the energy statists. They just turned the argument upside-down after the mid-1980s oil glut rendered their fossil fuel *exhaustion* theory a laughingstock. So by the late 1980s, they were claiming, in opposite fashion, that the world was actually suffering from a *surfeit* of fossil fuel owing to global warming!

Needless to say, the so-called AGW (anthropogenic global warming) threat is based on junk science and the will to power by the UN and national government statists who make a living and career out of fighting the fossil fuel industry. As is evident from the geology-based graph below, however, both current global temperatures and CO2 concentrations at 15 C and 440 ppm, respectively, ***are at epochal lows.*** Especially during the most recent 2.6 million year "ice age" both measures have fallen to below levels prevalent 90% of the time during the last 600 million years of the modern planetary climate.

Moreover, the counterclaim that this time is different because 7 billion humans didn't inhabit the planet during the vast stretches of geologic time shown below–with far higher CO2 levels and temperatures–is a *non sequitur.* That's because the global warming alarmists' claim is actually about planetary climate mechanics apart from what areas of the globe are currently populated by greenhouse gas-emitting homo sapiens. Supposedly, higher CO2 levels will cause the earth's climate to succumb to a doom loop–an irreversible and self-fueling heating cycle of ever higher CO2 levels and temperatures which will eventually cause all planetary life to become unsustainable.

Except. Except. That particular science experiment was run several times by Mother Nature during the last 600 million years and no doom loop ever happened. Instead, countervailing and self-correcting forces have caused the climate lows to rise and the climate highs to fall.

Accordingly, a further unwinding of the actual and natural climate change process that has been underway for 4.6 billion years is no existential threat at all, but in this instance is actually a blessing. That's because all things considered marginally higher CO2 levels and temperatures actually function to green the planet, enhance food production, reduce the hardships of colder climates, and elevate economic production and living standards for all of mankind.

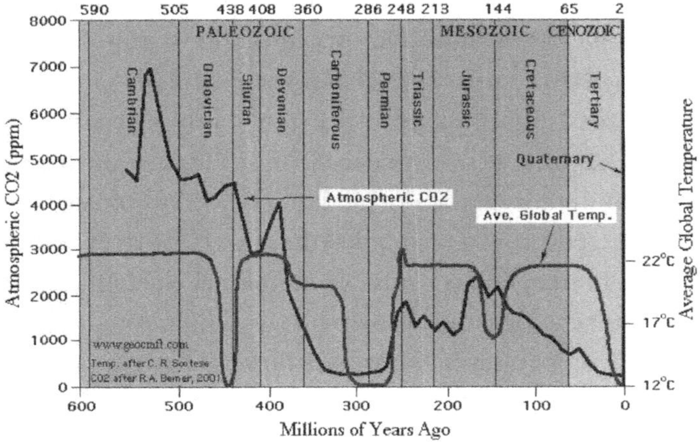

In short, the energy sector is tailor-made for free market economics, just like most other areas of modern economic life. So every vestige of Washington "energy policy" is a detriment to not just economic freedom, but to economic growth, rising living standards, and societal opportunity and fairness.

Accordingly, all forms of Federal intervention in the energy sector amount to government-fostered fraud, waste, and abuse. On the regulatory side that means repeal of the CAFE standards for autos and a whole host of energy efficiency standards for appliances ranging from toasters to refrigerators and gas furnaces, as well as canceling the outright regulatory war on coal and other fossil-fired utility plants.

Likewise, all the DOE energy spending and energy tax expenditure programs should be defunded, thereby generating budget savings of upwards of $60 billion by FY 2029. Thus, the tax expenditures itemized below unnecessarily subsidize fossil fuels and green energy alike (items # 1-5), while the DOE direct spending subsidies (items #6-15) are aimed at artificially and counterproductively shifting the mix of energy flavors from lower-cost fossil fuels to higher-cost green energy in furtherance of the net zero scam.

FY 2029 Savings From Defunding Energy Fraud, Waste and Abuse Subsidies:

1. Oil, Gas, and Coal depletion and E&D expensing allowances: $1.6 billion.
2. Enhanced Oil Recovery and Stripper Well Credits: $1.9 billion.
3. Green Energy Production and Investment Credits: $9.6 billion.
4. Carbon Sequestration Tax Credits: $3.3 billion.
5. Other Energy Tax Expenditures: $1.3 billion.
6. DOE Renewable Energy Integration Subsidies: $4.0 billion.
7. Advanced Reactor Demonstrations and other nuclear subsidies: $4.5 billion.
8. Clean Hydrogen Projects: $2.0 billion.
9. Energy Storage Demonstrations: $1.4 billion.
10. Grid Modernization and Resilience Projects: $7.2 billion.
11. State and Local Energy Projects: $2.5 billion.
12. Other DOE Clean Energy programs: $3.2 billion.
13. DOE Solar, Wind, and Renewable Projects: $10.0 billion.

14. Weatherization and Home Retrofit Subsidies: $6.5 billion.
15. Carbon Capture Subsidies: $1.0 billion.
16. Total Defunded Energy Subsidies: ***$60.0 billion.***

Other Business Special Interest Waste

The easiest place to start is with the $1.8 billion annual budget of the Small Business Administration (SBA). The whole purpose of the agency is to divert money to notionally small businesses that purportedly can't get funded in the private market. But the presumption that an investable business can't access funding in today's massively financialized economy is veritable hogwash. After all, at the present time there are 4,850 banks, 550 savings and loans, 1,200 nonbank lenders, and 2,200 venture capital funds in the US–all hungry for customers and opportunities to put their capital to work.

Moreover, according to a Federal Reserve Survey of small businesses, banks alone originated upwards of $600 billion in new loans to small businesses with revenues under $10 million in 2023. At the same time, VCs invest about $175 billion per year–almost all of which goes to start-up and small businesses. So there is absolutely no reason for Washington to be in the small business loan and loan guarantee business. To wit, if wanna-be small businesses can't get loans or venture investments from the huge available private financing market, they are likely poor risks. And if what they really want is below market rates and terms, why is it the taxpayers' job to foot the implicit subsidy?

As it happens, the SBA funds or guarantees about ***50,000*** loans per year with disbursements of about ***$20 billion*** under its main Section 7(a) loan guarantee program. But that's a drop in the bucket compared to the size of the ***$800 billion*** small business financing market referenced above or relative to the small business sector generally.

As shown below, there are approximately 15.2 million small businesses in the US with revenues under $10 million, meaning that at best SBA serves about *0.3%* of the relevant universe. In fact, nearly all of the businesses in the first two lines of the table below would meet the SBA size limits for eligibility, where cutoff points range from $7 million of annual sales for applicants in the retail sector to $8 million for manufacturing and agriculture, $21 million for wholesale trade, and $36 million for construction industry businesses.

Yet somehow 99%+ of these 15 million small businesses manage to finance themselves by means of internal cash flows or outside loans and capital without any help from the SBA whatsoever. So if the SBA disappeared tomorrow, the huge, booming small business sector of the US economy depicted in the table below would not miss a beat.

The fact that the Federal government maintains an unneeded agency employing 3,500 bureaucrats is a testament to the lobby muscle of the small business organizations and the members of the small business committees of the House and Senate, and nothing more. But if there is zero state interest in the SBA's operations and mission, then its budget does indeed constitute fat and waste in every sense of the word. Savings of direct budget outlays of $1.5 billion per year could be generated by abolishing the SBA.

Distribution of US Businesses By Size

Revenue Size Category	Revenue Range	Number of Businesses	Percentage of Total Businesses
Micro Businesses	Less than $1 million	14,382,087	90.4%
Small Businesses	$1 million to $10 million	818,345	5.1%
Medium-Sized Businesses	$10 million to $100 million	559,315	3.5%
Large Businesses	$100 million to $1 billion	148,583	0.9%
Enterprise Businesses	Over $1 billion	5,334	0.03%
Total		**15,913,664**	100%

What is true for the small business sector is true in spades when it comes to the Export-Import Bank (EXIM) which provides subsidized loan guarantees to US exporters. Most of the financing goes to giant companies like Boeing and Caterpillar, which are more than capable of generating private sector funding for their exports. So what EXIM actually does is put taxpayers in the equation in order to ***subsidize*** the export financing cost for projects like the one described below, which was typical among the $7 billion of financings done by EXIM in 2023.

We are referring to a $594 million loan guarantee to Kazakhstan Temir Zholy (KTZ), the national railway of Kazakhstan, to finance the export of Wabtec Corporation's locomotive and critical rail equipment to KTZ. The transaction purportedly supported an estimated 1,500 U.S. jobs at Wabtec in the Western Pennsylvania region.

Yet what unadulterated tommyrot. Wabtec Corporation is a $10 billion company with $1 billion of operating income, total assets of $18.5 billion, equity of $10 billion, and 27,000 employees. It does export business in 50 countries out of manufacturing operations based in the U.S., Europe, Canada, Mexico, Australia, and South America.

Without any help from Uncle Sam at all, therefore, those

1,500 jobs would exist in support of Wabtec's ample global export business. The only difference, of course, is that the interest rate on the above export loan, if it were not financed by Uncle Sucker, might be say 200 basis points higher–meaning that it would reduce Wabtec's after-tax net income by about $9 million.

So what! That would amount to just 1.5% of Wabtec's 2022 net income of $641 million. Thus, terminating the $400 million annual cost of the EXIM's payroll and loss costs is a no-brainer, if there ever was one.

In short, the Federal budget is larded with subventions and corporate welfare that supposedly support "jobs," but actually simply fattens business profits as a practical matter. In the case of SBA and EXIM alone, abolition of these crony capitalist agencies could save ***$2 billion*** per year. But that's only the tip of the iceberg. For instance, the budgets for the Forest Service, the Bureau of Land Management, the Bureau of Reclamation, and the Army Corps of Engineers totaled about $23 billion on a combined basis in FY 2023.

While the rationale for these expenditures is that they promote local economies, jobs, and public amenities, the fact is that the spillover gains to businesses benefiting from their operations and projects are enormous. Very easily, therefore, an appropriate increase in user fees and assessments on businesses capturing windfalls from these public works activities could total 20% of outlays or upwards of ***$4 billion*** per year in Federal budget savings.

Another opportunity to capture windfall waste lies in the out-of-control funding of the National Institutes of Health (NIH). Not only have these agencies become increasingly captive of Big Pharma and Big Health Care, but their budgets have soared far beyond any reasonable level. For instance, the already ample NIH budget in the year 2000 was about $20 billion, meaning that indexed for inflation it should have totaled about $34 billion by FY 2023.

Alas, the actual NIH budget last year weighed in at $48 billion! This means its inflation-adjusted budget has risen by 42% since the turn of the century–even as the funding priorities of the NIH have shifted massively toward commercially-oriented research and development of drugs and other medical therapies that should be financed by these already profit-oozing industries. Accordingly, a ***$14 billion*** cutback to a tighter scientific basics approach and to the more than ample inflation-adjusted level of the year 2000 is fully warranted.

However, the greatest area for eliminating business subsidies and boondoggles is in the tax code. While we think taxes should always be lower rather than higher–so long as budgets are balanced–the practice of muscling into the tax code loopholes, incentives, and deductions for special interest groups and industries is every bit as wasteful as unneeded direct spending.

As it happens, the US Treasury publishes a periodic report on "tax expenditures." These estimates basically measure special interest provisions of the tax code by comparing the actual revenues generated by these provisions versus what would occur on a level playing field in which all taxpayers were assessed the actual statutory rates currently embedded in the IRS code. And these tax expenditure estimates are also projected through our target budget year of FY 2029.

On that basis, there are nine areas where the IRS code generates huge windfalls to businesses and investors. These are listed below and for the most part are self-explanatory. For instance, the huge $55 billion tax expenditure for foreign source income is owing to the fact that US citizens living and working abroad get a $126,000 deduction from US taxable income; qualifying US exporters pay a rate of just 16% versus the statutory 21% on corporate income; and US corporations operating abroad pay essentially a rate of just 13% on foreign source income.

Similarly, the $34 billion value of the R&D credit is just a function of the lobbying clout of the tech sector. There is no tax policy or economic reason why R&D expenses should get more favorable treatment from the tax code than spending for advertising, staff training, machinery maintenance, or financial planning. Costs are costs until lobbying clout says otherwise.

Likewise, the $28 billion of loopholes and incentives for business investment in housing are exceedingly inefficient and mistargeted. If there is a case to subsidize the housing expenses of low-income and/or renter households, it should be done through direct payments to eligible populations, as specified by Congressionally-approved and funded income transfer programs–not via windfalls to subsidized investors.

The $27 billion per year of combined costs for the financial institutions tax preference, investments in computer equipment and software of up to $1 million, new investments in poverty zones, and the like-kind exchange exemption from capital gains tax are all pure products of the so-called Gucci Gulch–the army of K-Street lobbyists which forms outside the Ways and Means Committee room at tax writing time. There is no economic basis for any of these subventions relative to the otherwise applicable statutory rates.

In all, if the IRS code operated on a level playing field with respect to taxation of business activities and corporate income, Federal revenue collections would be ***$150 billion*** per year higher by our target year of 2029.

Pro-Business Tax Expenditures To Be Defunded:

- $126,000 deduction for Income Earned Abroad: $9.7 billion.
- Lower Rates on Export Earnings: $10.0 billion.
- Other tax preference for Foreign Source Business income: $35.4 billion.
- R&D Credit: $34.4 billion.
- Agriculture, Natural Resources and Timber preferences: $5.4 billion.
- Housing investment incentives: $27.9 billion.
- Financial institution tax preferences: $8.3 billion.
- Expensing of small computer investments: $12.5 billion.
- Like-kind exchange preference and New Markets Tax credit: $6.4 billion.
- Total, Pro-Business Tax Expenditures: ***$150.0 billion.***

In all, savings from eliminating business subsidies and tax preferences would amount to $170 billion by the FY 2029 target year. That's no inconsiderable sum, but this exercise does underscore that achieving the necessary $2 trillion per year of Federal budget savings will be no cakewalk. Even after aggressively slashing fat in the Federal bureaucracy and terminating wasteful subsidies to a sweeping range of the most powerful Washington lobbies, as summarized by the table below, the savings would only total ***$400 billion*** or just *20%* of the $2 trillion target.

What that means, of course, is you can't even begin to stem the hemorrhaging public debt by cutting fat, waste, and bureaucracy alone. As we show in the next sections, therefore, you have to take huge chunks out of the unnecessarily muscled national security budget and even larger extractions from the bone of Federal entitlement spending.

Summary of FY 2029 Savings From Slashing the Fat:

- Agency Headcount Cuts and Related Savings: $130.0 billion.
- Farm Subsidy Eliminations: $40.0 billion.
- Energy Industry Subsidy Eliminations: $60.0 billion.
- Business Subsidy Cuts: $170.0 billion.
- Total, Slash the Fat Savings: ***$400 billion.***

CHAPTER 7

America First–The Key To A $500 Billion Shrinkage of The Warfare State's Excess Muscle

If Donald Trump's "America First"-focused foreign policy means anything at all, it's that the current $1 trillion national security budget is *double the size* that an adequate homeland defense shield actually requires. Indeed, it is no exaggeration to say that in relentless pursuit of its own self-serving aggrandizement, the military/industrial/intelligence complex has massively inflated America's Warfare State into an "extra-large" when what is really needed in the world of 2024 is a snug-fitting "small."

The basis for that stunning disconnect goes back deep into Cold War history and its aftermath. The post-WWII policy of collective security, extensive alliances through NATO and its regional clones and globe-spanning military power projection capabilities, and a network of 750 foreign bases was an epic historical mistake. It fostered the opposite of America First and permanently broke faith with Thomas Jefferson's wise admonition urging,

"...peace, commerce, and honest friendship with all nations, ***entangling alliances with none.***"

At length, Washington became the War Capital of the World and the seat of an Empire First policy regime embraced by both elected officialdom and the multitudinous nomenklatura of the Warfare State that took up permanent residence on the banks of the Potomac. In fact, the Empire First policy regime became so deeply rooted that even 34 years after the Soviet Union disappeared into the dustbin of history, America's elephantine Warfare State refuses to go quietly into the good night.

The reason, of course, is that today's Warfare State never was grounded in an objective external threat. Even during Soviet times, the exaggerated girth of America's military machine was based on vast threat inflations emanating from a resource-heavy national security bureaucracy seeking to secure its own future funding and to relentlessly expand its missions and remit.

That Washington's trillion-dollar Warfare State is rooted in internal self-perpetuation rather than external threats is evident from the post-Cold War dog that didn't bark. That is, the Soviet archives are now open, but there's absolutely nothing there to validate the Cold War axiom that the Soviet Union–along with the affiliated menace of Maoist China–was hellbent on world military domination, starting with Western Europe, Japan, and then extending to the lesser lands all around them.

In fact, the Soviet archives make clear that Moscow never had a plan or even faint aspiration to fortify and offensively unleash the Red Army toward Bonn, Paris, and London. The closest thing to a plan for military mobilization westward was the "Seven Days to the Rhine" blueprint, but that was a defensive action plan explicitly formulated as a contingency response to a theoretical NATO first strike.

According to the plan, if NATO were to launch a nuclear attack on Poland, the Warsaw Pact would respond with a massive counterattack aimed at quickly overwhelming NATO forces in Western Europe. The goal was to reach the Rhine River within seven days, effectively splitting Europe and preventing NATO reinforcements from reaching the front lines in Eastern Europe and potentially embarking upon yet a fourth post-1800 invasion of Mother Russia.

Indeed, what the Soviet archives actually show is not the deliberations of a menacing military Colossus, but the record of a chronic struggle to hold together with economic baling wire and bubble gum a lumbering communist state that didn't function and couldn't last.

Nevertheless, it was the false fear of a red tide descending over Europe and ultimately the Western Hemisphere, too, that enabled Empire First to trump the natural and proper tendency of Washington politicians and policy-makers to retreat behind America's secure ocean moats after WWII. In fact, for a brief interlude, a sweeping military demobilization did occur, when the peak $83 billion defense budget of 1945 plunged to just $9 billion by 1948.

But that sensible attempt for the second time in the 20th century at post-war demobilization and a return to peacetime normalcy was reversed in 1949 when the Soviet Union got the A-bomb, and Mao won the civil war in China. Thereafter, the spread of bases, troops, alliances, interventions, and Forever Wars proceeded relentlessly on the grounds that the rickety communist states domiciled in Moscow and Beijing posed an existential threat to America's survival.

They did not. Not by a long shot. As the great Senator Robert Taft held at the time, the modest threat to homeland security presented by the war-ravaged corpus of the Soviet Union and the collectivist disaster imposed on China by Mao could have been readily handled with–

- An overwhelming strategic nuclear retaliatory capacity that would have deterred any possibility of nuclear attack or blackmail.
- A Fortress America conventional defense of the continental shorelines and air space that would have been exceedingly easy to stand up, given that the Soviet Union had no Navy worth speaking of and China had devolved into industrial and agricultural anarchy owing to Mao's catastrophic experiments with collectivization.

That eminently correct Taftian framework never did change through the end of the Cold War in 1991, even as the technology of nuclear and conventional warfare evolved apace. For modest military spending, Washington could have kept its nuclear deterrent fully effective and maintained a formidable Fortress America defense of the homeland without any of the apparatus of Empire and no American boots on foreign soil, at all. And after 1991, the requirement would have been even less demanding.

In fact, the case for a true America First policy–that is, returning to the 1948 status quo ante and a proper Fortress America military posture–has powerfully strengthened during the last three decades. That's because, in today's world, the only theoretical military threat to America's homeland security is the possibility of nuclear attack or blackmail. That is to say, the threat of an adversary with a First Strike capacity so overwhelming, lethal, and effective that it could simply call out checkmate and demand Washington's surrender.

Fortunately, there is no nation on earth that has anything close to the Nuclear First Strike force that would be needed to totally overwhelm America's triad nuclear deterrent and thereby avoid a retaliatory annihilation of its own country and people if it attempted to strike first. After all, the US has 3,700 active nuclear warheads, of which about 1,800 are operational at any point in time. In turn, these are spread under the seven seas, in hardened silos, and among a bomber fleet of 66 B-2 and B-52s–all beyond the detection or reach of any other nuclear power.

For instance, the Ohio class nuclear submarines each have 20 missile tubes, with each missile carrying an average of 4-5 warheads. That's 90 independently targetable warheads per boat. At any given time 12 of the 14 Ohio class nuclear subs are actively deployed and spread around the oceans of the planet within a firing range of 4,000 miles.

So at the point of attack, that's ***1,080 deep-sea nuclear warheads*** cruising along the ocean bottoms that would need to be identified, located, and neutralized before any would-be nuclear attacker or blackmailer even gets started. Indeed, with respect to the "Where's Waldo?" aspect of it, the sea-based nuclear force alone is a powerful guarantor of America's homeland security. Even Russia's vaunted hypersonic missiles couldn't find or take out by surprise the US sea-based deterrent.

And then there are the roughly 300 nukes aboard the 66

strategic bombers, which also are not sitting on a single airfield Pearl Harbor-style waiting to be obliterated either, but are constantly rotating in the air and on the move. Likewise, the 400 Minuteman III missiles are spread out in extremely hardened silos deep underground across a broad swath of the upper Midwest. Each missile currently carries one nuclear warhead in compliance with the Start Treaty but could be MIRV'd in response to a severe threat, thereby further compounding and complicating an adversary's First Strike calculus.

Needless to say, there is no way, shape, or form that America's nuclear deterrent can be neutralized by a blackmailer. And that gets us to the heart of the case for drastically downsizing America's military muscle. To wit, according to the most recent CBO estimates the nuclear triad will cost only about ***$75 billion per year*** to maintain over the next decade, including allowances for periodic weapons upgrades.

That's right. The core component of America's military security ***requires only 7% of today's massive military budget*** as detailed on a system-by-system basis in the table below. Thus, in 2023 the nuclear triad itself cost just $28 billion plus another $24 billion for related stockpiles and command, control, and warning infrastructure.

Moreover, the key component of this nuclear deterrent–the sea-based ballistic missile force–is estimated to cost just ***$188 billion*** over the *entire* next decade. That's only ***1.9%*** of the $10 trillion CBO defense baseline for that period.

10-Year Cost Of US Strategic Nuclear Deterrent Per CBO Estimates, 2023 to 2032

	2023			Total, 2023–2032		
	DoD	DOE	Total	DoD	DOE	Total
CBO's Projections of Budgeted Amounts for Nuclear Forces[a]						
Nuclear delivery systems and weapons						
Strategic nuclear delivery systems and weapons						
Ballistic missile submarines	11.4	1.2	12.7	172	16	188
Intercontinental ballistic missiles	6.4	0.9	7.3	103	16	118
Bombers	4.2	1.7	5.8	52	11	63
Other DoD nuclear activities[b]	1.6	n.a.	1.6	19	n.a.	19
Subtotal	23.6	3.8	27.5	346	43	389
Tactical nuclear delivery systems and weapons	0.6	0.4	1.0	5	2	6
Nuclear weapons laboratories and supporting activities						
Stockpile services	n.a.	1.1	1.1	n.a.	12	12
Facilities and infrastructure	n.a.	7.3	7.3	n.a.	79	79
Other stewardship and support activities[c]	n.a.	5.1	5.1	n.a.	57	57
Subtotal	n.a.	13.4	13.4	n.a.	148	148
Subtotal, Nuclear Delivery Systems and Weapons	24.2	17.7	41.9	351	192	543
Command, control, communications, and early-warning systems						
Command and control	1.5	n.a.	1.5	24	n.a.	24
Communications	2.7	n.a.	2.7	34	n.a.	34
Early-warning	6.3	n.a.	6.3	58	n.a.	58
Subtotal, Command, Control, Communications, and Early-Warning Systems	10.5	n.a.	10.5	117	n.a.	117
Total Budgeted Amounts for Nuclear Forces	**34.7**	**17.7**	**52.4**	**468**	**192**	**660**
CBO's Estimates of Additional Costs Based on Historical Cost Growth	n.a.	n.a.	n.a.	56	40	96
Total Estimated Cost of Nuclear Forces	**34.7**	**17.7**	**52.4**	**524**	**232**	**756**

So the question recurs with respect to the CBO's current $989 billion baseline spending level for defense in our target year of 2029. After setting aside $75 billion for the strategic nuclear triad, how much of the remaining $900 billion+ would actually be needed for a conventional Fortress America defense of the continental shorelines and airspace?

The starting point is that in the present world order, there are no technologically advanced industrial powers who have either the capability or intention to attack the American homeland with conventional forces. To do that you need a massive military armada including a Navy and Air Force many times the size of current US forces, huge air and sealift resources, and humongous supply lines and logistics capacities that have never been even dreamed of by any other nation on the planet.

You also need an initial GDP of say $50 trillion to sustain

what would be the most colossal mobilization of weaponry and materiel in human history. And that's to say nothing of needing to be ruled by suicidal leaders willing to risk the nuclear destruction of their own countries, allies, and economic commerce in order to accomplish, what? Occupy Denver?

The entire idea that there is a post-Cold War existential threat to America's security is just nuts. For one thing, nobody has the GDP or military heft. Russia's GDP is a scant $2 trillion, not the $50 trillion that would be needed for it to put invasionary forces on the New Jersey shores. And its defense budget is $75 billion, which amounts to about *four weeks of waste* in Washington's $900 billion monster.

As for China, it doesn't have the GDP heft to even think about landing on the California shores, notwithstanding Wall Street's endless kowtowing to the China Boom. The fact is, China has accumulated in excess of $50 trillion of debt in barely two decades!

Therefore, it didn't grow organically in the historic capitalist mode; it printed, borrowed, spent, and built like there was no tomorrow. The resulting simulacrum of prosperity would not last a year if its $3.6 trillion global export market--the source of the hard cash that keeps its Ponzi upright–were to crash, which is exactly what would happen if it tried to invade America.

To be sure, China's totalitarian leaders are immensely misguided and downright evil from the perspective of their oppressed population. But they are not stupid. They stay in power by keeping the people relatively fat and happy and would never risk bringing down what amounts to an economic house of cards that has not even a vague approximation in human history.

Indeed, when it comes to the threat of a conventional military invasion, the vast Atlantic and Pacific moats are even greater barriers to foreign military assault in the 21st century than they so successfully proved to be in the 19th century. That's because today's advanced surveillance technology and anti-ship missiles would consign an enemy naval armada to Davy Jones' Locker nearly as soon as it steamed out of its own territorial waters.

The fact is, in an age when the sky is flush with high-tech surveillance assets a massive conventional force armada couldn't possibly be secretly built, tested, and mustered for surprise attack without being noticed in Washington. There can be no repeat of the Japanese strike force–the *Akagi, Kaga, Soryu, Hiryu, Shokaku, and Zuikaku*–steaming across the Pacific toward Pearl Harbor sight unseen.

Indeed, America's ostensible "enemies" actually have no offensive or invasionary capacity at all. Russia has only one aircraft carrier–a 1980s-era relic that has been in dry dock for repairs since 2017 and is equipped with neither a phalanx of escort ships nor a suite of attack and fighter aircraft–and at the moment not even an active crew.

Likewise, China has just three aircraft carriers–two of which are refurbished rust buckets purchased from the remnants of the old Soviet Union (actually Ukraine!), and which carriers do not even have modern catapults for launching their strike aircraft.

In short, neither China nor Russia will be steaming their tiny 3- and 1-carrier battle groups toward the shores of either California or New Jersey anytime soon. An invasionary force that had any chance at all of surviving a US fortress defense of cruise missiles, drones, jet fighters, attack submarines, and electronics warfare would need to be 100X larger.

Again, there is also no GDP in the world–$2 trillion for Russia or $18 trillion for China–that is even remotely close in size to the $50 trillion, or even $100 trillion, that would be needed to support such an invasionary force without capsizing the home economy.

Yet and yet. Washington still maintains a globe-spanning conventional war-fighting capability that it never really needed even during the Cold War. But now, fully one-third of a century after the Soviet Empire collapsed and China went the Red Capitalist route of deep global economic integration, it amounts to utterly extraneous and unneeded muscle.

We are referring, of course, to the 173,000 US troops in 159 countries and the network of 750 bases in 80 countries. All told, Washington equips, trains, and deploys an armed force of 2.86 million not for purposes of homeland defense but overwhelmingly for missions of overseas offense, invasion, and occupation all over the planet.

As depicted in the graphic below, this obsolete Empire First military posture still includes among others–

- 119 facilities and nearly 34,000 troops in Germany.
- 44 facilities and 12,250 troops in Italy.
- 25 facilities and 9,275 troops in the UK.
- 120 facilities and 53,700 troops in Japan.
- 73 facilities and 26,400 troops in South Korea

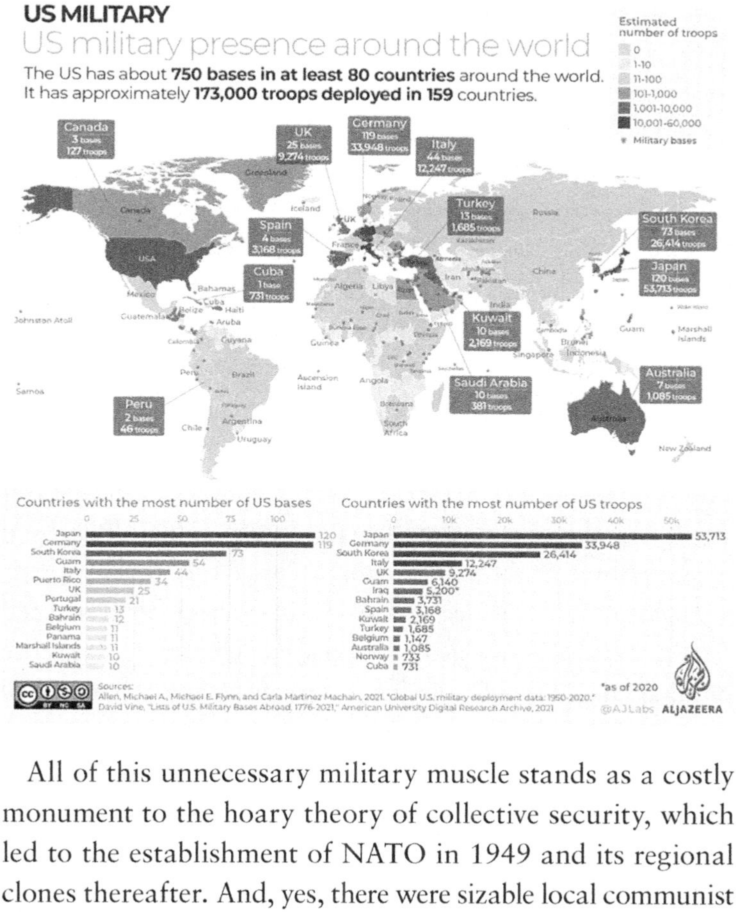

All of this unnecessary military muscle stands as a costly monument to the hoary theory of collective security, which led to the establishment of NATO in 1949 and its regional clones thereafter. And, yes, there were sizable local communist parties in Italy and France in the late 1940s, and the Labor Party in England had a reddish hue. But, again, the now open archives of the old Soviet Union prove conclusively that Stalin had neither the wherewithal nor intention to invade Western Europe.

What military capacity the Soviet Union did resurrect after the bloodletting with Hitler's armies was heavily defensive in character and lumbering in capabilities. So the communist threat in Europe could have been wrangled out by these nations at the polls, not on the battlefield. They did not need NATO to stop an imminent Soviet invasion.

Of course, what NATO did accomplish was to reduce dramatically the burden of defense spending in Western Europe, even as most of these nations opted for an expansive and expensive Welfare State. That is to say, the bloated Warfare State that America didn't need from 1950 through 1990 ultimately enabled the Welfare States that Europe couldn't afford, either then or now.

Needless to say, once the Washington-based Empire of bases, alliances, collective security, and relentless CIA meddling in the internal affairs of foreign countries was established, it stuck like glue–even as the facts of international life proved over and over again that the Empire wasn't needed.

That is to say, the alleged "lessons" of the interwar period and WWII were falsely played and replayed. The aberrational rise of Hitler and Stalin did not happen because the good people of England, France, and America slept through the 1920s and 1930s.

Instead, they arose from the ashes of Woodrow Wilson's pointless intervention in a quarrel of the Old World that was none of America's business. Yet the arrival in 1918 of two million American doughboys and massive flows of armaments and loans from Washington enabled a vindictive peace of the victors at Versailles rather than an end to a desultory world war that would have left all the sides exhausted, bankrupt, and demoralized, and their respective domestic war parties subject to massive repudiation at the polls.

As it happened, however, Wilson's intervention on the stalemated battlefields of the Western Front gave birth to

Lenin and Stalin, while his machinations with the victors at Versailles fostered the rise of Hitler.

To be sure, in the end the former did fortunately bring about the demise of the latter at Stalingrad. But that should have been the end of the matter in 1945, and, in fact, the world was almost there. After the victory parades, demobilization and normalization of civilian life proceeded apace all around the world.

Alas, Washington's incipient War Party of military contractors and globetrotting operatives and officialdom gestated in the heat of World War II was not about to go quietly into the good night. Instead, the Cold War was midwifed on the banks of the Potomac when President Truman fell under the spell of war hawks like Secretary James Byrnes, Dean Acheson, James Forrestal, and the Dulles brothers, who were loath to go back to their mundane lives as civilian bankers, politicians, or peacetime diplomats.

So, in the post-war period world communism was not really on the march and the nations of the world were not implicated in falling dominoes, nor were they gestating incipient Hitlers and Stalins. But the new proponents of Empire insisted they were just the same, and that national security required the far-flung empire that is still with us today.

So there is no mystery, therefore, as to why the Forever Wars go on endlessly. Or why at a time when Uncle Sam is hemorrhaging red ink like never before, a large bipartisan majority sees fit to authorize $1.1 trillion per year for vastly excessive military muscle and wasteful foreign aid boondoggles that do absolutely nothing for America's homeland security.

In effect, Washington has morphed into a freak of world history–a planetary War Capital dominated by a panoptic complex of arms merchants, paladins of foreign intervention and adventure, and Warfare State nomenklatura. Never before has there been assembled and concentrated under a single state authority a hegemonic force possessing such enormous fiscal resources and military wherewithal.

Not surprisingly, the War Capital on the Potomac is Orwellian to the core. War is always and everywhere described as the promotion of peace. Its jackboot of global hegemony is gussied up in the beneficent-appearing form of alliances and treaties. These are ostensibly designed to promote a "rules-based order" and collective security for the benefit of mankind, not simply the proper goals of peace, liberty, safety, and prosperity within America's homeland.

As we have seen, however, the whole intellectual foundation of this enterprise is false. The planet is not crawling with all-powerful would-be aggressors and empire-builders who must be stopped cold at their own borders, lest they devour the freedom of all their neighbors near and far.

Nor is the DNA of nations perennially infected with incipient butchers and tyrants like Hitler and Stalin. They were one-time accidents in history and fully distinguishable from the standard run of everyday tinpots which actually do arise periodically. But the latter mainly disturb the equipoise of their immediate neighborhoods, not the peace of the planet.

So America's homeland security does not depend upon a far-flung array of alliances, treaties, military bases, and foreign influence operations. In today's world, there are no Hitlers, actual or latent, to stop. The whole framework of Pax Americana and the Washington-based promotion and enforcement of a "rules-based" international order is an epic blunder.

In that regard, the Founding Fathers got it right more than 200 years ago during the infancy of the Republic. As John Quincy Adams approvingly held,

> *"[America] has abstained from interference in the concerns of others, even when conflict has been for principles to which she clings...She is the well-wisher to the freedom and independence of all. She is the champion and vindicator only of her own."*

Needless to say, peaceful commerce is invariably far more beneficial to nations large and small than meddling, interventionism, and military engagement. In today's world it would be the default state of play on the international chessboard, save for the Great Hegemon on the banks of the Potomac. That is to say, the main disturbance of the peace today is invariably fostered by the self-appointed peacemaker, who, ironically, is inherently the least threatened large nation on the entire planet.

The starting point for an America First military posture, therefore, is the drastic downsizing of the nearly one-million-man standing US Army. The latter would have no uses abroad because there would be no cause for wars of foreign invasion and occupation, while the odds of any foreign battalions and divisions reaching America's shores are virtually non-existent. With a proper coastline garrison of missiles, drones, attack submarines, and jet fighters any invading army would become shark bait long before it saw the shores of California or New Jersey.

Yet the 462,000 active-duty army soldiers at $112,000 each have an annual budget cost of $55 billion while the 506,000 army reserve forces at $32,000 each cost upwards of $16 billion. And on top of this force structure, of course, you have $77 billion for operations and maintenance, $27 billion for procurement, $22 billion for RDT&E, and $4 billion for everything else (based on the FY 2025 budget request).

In all, the current Army budget totals nearly ***$200 billion***, and virtually all of that massive expenditure--nearly 3X the total defense budget of Russia–is deployed in the service of Empire, not homeland defense. It could readily be cut by 70% or $140 billion–meaning that the US Army component of a $450 billion Fortress America defense budget would absorb just $60 billion annually.

Likewise, the US Navy and Marine Corps spend $55 billion annually on 515,000 active-duty forces and another $3.7 billion on 88,000 reserves. Yet if you look at the core requirements of a Fortress America defense posture, these forces and expenses are way over the top, as well.

By core missions, we refer to the Navy component of the strategic nuclear triad and the Navy's large force of attack and cruise missile submarines. As it happens, here are the current manpower requirements for these key forces:

- ***14 Ohio-class Strategic Nuclear Subs:*** There are two crews of 155 officers and enlisted men for each boat, resulting in a direct force requirement of 4,400 and an overall total of 10,000 military personnel when Admirals, overhead, support, and woke compliance is included (or not).
- ***@50 Attack/Cruise Missile Subs:*** There are two crews of 132 officers and enlisted men for each boat, for a direct requirement of 13,000 and an overall total of 20,000 including Admirals and overhead, etc.

In short, the core Navy missions of a Fortress America defense involve about 30,000 officers and enlisted men, or less than *6%* of the current active-duty force of the Navy/Marine Corps. On the other hand, the totally unnecessary carrier battle groups, which operate exclusively in the service of Empire, have crews of 8,000 each when you count the escort ships and suites of aircraft.

So, the 11 carrier battle groups and their infrastructure require *88,000* direct military personnel and ***140,000*** overall when you include the usual support and overhead. Likewise, the active-duty force of the Marine Corps is *175,000,* and that's entirely an instrument of invasion and occupation. It's totally unnecessary for a homeland defense.

In short, fully ***315,000 or 60%*** of the current active-duty force of the Navy/Marine Corps functions in the service of Empire. So, if you redefine the Navy's missions to focus on strategic nuclear deterrence and coastal defense, it is evident that more than half of the Navy's force structure is not necessary for homeland security. Instead, it functions in the service of global power projection, policing of the sea lanes from the Red Sea to the East China Sea and platforming for wars of invasion and occupation.

Overall, the current Navy/Marine Corps budget stands at about ***$236 billion*** when you include $59 billion for military personnel, $81 billion for O&M, $67 billion for procurement, $26 billion for RDT&E, and $4 billion for all others. A $96 billion or ***40% cut,*** therefore, would still leave ***$140 billion*** for the core missions of a Fortress America defense.

Among the services, the $246 billion contained in the Air Force budget is considerably more heavily oriented to a Fortress

America versus Empire-based national security posture than is the case with the Army and Navy. Both the Minuteman land-based leg of the strategic triad and the B-52 and B-2 bomber forces are funded in this section of the defense budget.

And while a significant fraction of the budget for the manning, operations, and procurement of conventional aircraft and missile forces is currently devoted to overseas missions, only the airlift and foreign base component of those outlays inherently function in the service of Empire.

Under a Fortress America defense, therefore, a substantial part of the conventional air power, which includes upwards of 4,000 fixed-wing and rotary aircraft, would be repurposed to homeland defense missions. Accordingly, upwards of 75% or $180 billion of the current Air Force budget would remain in place, limiting the savings to just $65 billion.

Finally, an especially sharp knife would be brought down upon the ***$181 billion*** component of the defense budget which is for the Pentagon and DOD-wide overhead operations. Fully ***$110 billion*** or 61% of that huge sum–again more than 2X the total military budget of Russia–is actually for the army of DOD civilian employees and DC/Virginia based contractors which feast upon the Warfare State.

In terms of homeland security, much of these expenditures are not simply unnecessary--they are actually counter-productive. They constitute the taxpayer-funded lobby and influence-peddling force that keeps the Empire alive and fully funded on Capitol Hill. Even then, a 38% allowance or $70 billion for the Defense Department functions (which include massive overfunding of the CIA and other intelligence agencies) would more than provide for the true needs of a

Fortress America defense.

Overall, therefore, downsizing the DOD muscle would generate $410 billion of savings on an FY 2025 basis. Another $50 billion in savings could also be obtained from eliminating most funding for the UN, other international agencies, security assistance, and economic aid. Adjusted for inflation through 2029 the total savings would come to $500 billion.

Fortress America Budget Savings:

- Army: $140 billion.
- Navy/Marine Corps: $96 billion.
- Air Force: $65 billion.
- DOD agency-wide: $111 billion.
- UN contributions and foreign economic and humanitarian aid: $35 billion.
- International Security Assistance: $15 billion.
- Total Savings, FY 2025 basis: $462 billion.
- 8% inflation adjustment to FY 2029: +$38 billion.
- Total FY 2029 Budget Savings: ***$500 billion.***

The resulting allowances (FY 2025 basis) of $60 billion for the Army, $140 billion for the Navy, $180 billion for the Air Force, and $70 for DOD-wide operations would shrink the defense component of the Warfare State to $450 billion per year. In current dollars of purchasing power that happens to be exactly what Eisenhower thought was more than adequate for national security when he warned of the military-industrial complex during his farewell address 64 years ago.

At the end of the day, the time to bring the Empire home is

long overdue. The $1.3 trillion annual cost of the Warfare State (including international operations and veterans) is no longer even remotely affordable--and it has been wholly unnecessary for homeland security all along.

CHAPTER 8

How To Cut The Welfare State Entitlement Bone By $1.1 Trillion

After "Slashing the Fat" by upwards of 535,000 jobs or 35% of the nondefense Federal bureaucracy and downsizing the muscle of Washington's vastly bloated Warfare State by 50%, the total budget savings by FY 2029 would add up to a creditable ***$900 billion.*** But that's not nearly enough to reach the $2 trillion per year DOGE target, and also far short of what would be required to actually do the job of fiscal stabilization.

The fact is, there is absolutely no way to reach the $2 trillion target without cutting deeply into the bone of entitlement spending. And some kind of Brobdingnagian bone it is, as is evident from the CBO baseline numbers for the next decade (2025 to 2034).

To wit, even with the help of Rosy Scenario economic assumptions, CBO projects Federal revenues of $62.8 trillion over the next 10-year budget window while spending would actually total $84.9 trillion, leaving the aforementioned $22 trillion torrent of red ink. So even under CBO's optimistic assumption of full-employment revenues for the entire coming decade, the budget equation cannot be stabilized at current levels of taxation without sweeping reductions in the $85 trillion of baseline spending.

That gets us to the "bone," of course, because all $900 billion of annual savings (FY 2029) so far have come from the so-called discretionary accounts on the DOD and domestic programs sides of the Potomac, respectively. Yet these so-called discretionaries amount to just ***$20.5 trillion*** or 24% of the CBO spending baseline over the next decade. The ***$64 trillion*** remainder of the $85 billion total is owing to net interest and entitlements/mandatories.

Needless to say, the $13 trillion portion of the latter figure, which is attributable to net interest, has to be paid lest the entire debt-entombed US financial system implode. So the "bone" of entitlements/mandatories, which amounts to ***$51.4 trillion*** over the next decade, is the actual target for the huge amount of additional budget savings needed to reach the DOGE target.

On the surface, that figure may sound impressive in terms of the implied ample opportunities for savings, and by FY 2029 entitlements/mandatories would indeed amount to a hefty $4.86 trillion of annual spending. Still, the remaining $1.1 trillion of savings to hit the DOGE target would amount to paring the "bone" by a thumping ***23%***.

Furthermore, the composition of this $4.86 trillion bone only further underscores the severity of the challenge. To wit, fully $3.43 trillion of 71% of the CBO's baseline spending for entitlements/mandatories in FY 2029 consists of *Social Security and Medicare outlays*!

Yet whether this is the third rail of American politics or not, a considerable share of the bone-cutting would have to be done in these programs if the numbers have any chance of parsing at all.

As a practical matter, however, there is a numerically feasible and fully justifiable way to shrink these enormous Social Security/Medicare outlays: Namely, to means-test the millions of upper-income beneficiaries who do not need these benefits and who did not "earn" them on an actuarial basis anyway, and to also subject the Medicare reimbursement system to sweeping efficiency improvements.

However, to calibrate exactly how deep these means-tested cuts and efficiency improvements need to go, we back into the number by estimating savings that could be attained via maximum feasible cuts in all other entitlement areas of the budget that have not been addressed thus far.

These savings for FY 2029 are summarized in the table below and are then explained and amplified in the sections which follow. The key point here, however, is that you can't get to the $1.1 trillion cut-the-bone savings target by 2029 without upwards of ***21% or $725 billion*** of reductions from the $3.44 trillion baseline for Social Security and Medicare.

Summary if FY 2029 DOGE Cut-The-Bone Savings:

- Block Grant current means-tested entitlements and cut by 25%: $290 billion.
- Block Grant education and social services programs and cut by 30%: $35 billion.
- Block Grant and cut student aid by 40%: $50 billion.

- Improve Medicare efficiency by 10%: $150 billion.
- Means-test Social Security and Medicare, saving 17.5% of net outlays: $575 billion.
- Total Savings From Cutting-the-Bone: ***$1.1 trillion.***

25% Safety Net Savings

We refer to the current means-tested benefits programs as the Little Welfare State. The latter generated Federal outlays of $983 billion in FY 2024 and will grow to $1.15 trillion by FY 2029. The basis for the 25% cut shown in the summary above is that these programs never should have been Federal responsibilities in the first place. They are the legacy of the New Deal, Great Society, and a fair amount of monkeying around by the Nixon Administration and subsequent GOP presidents, save for Ronald Reagan.

More importantly, means-tested welfare simply does not lend itself to one-size-fits-all Federal standards and transfer payment levels owing to–

- substantial cost-of-living difference around the country.
- Wide differences in red versus blue state policy viewpoints when it comes to taxing producers to support non-producers.
- Valid design arguments about the mix of a cash versus an in-kind benefit approach to welfare.

In the latter case, for example, we are referring to in-kind help through Medicaid or subsidized housing units versus direct cash transfers like family assistance, SSI, and quasi-cash transfers such as food stamps. More cash and fewer in-kind benefits, of course, enable more recipient choices and responsibility, on the one hand, while in-kind benefits are purportedly less subject to abuse and irresponsibility, on the other.

After 50 years of experience, however, it is evident that Washington has no monopoly on the right answers to these nettlesome policy questions; and that the further away from local taxpayers and producers these transfer payment policy decisions lie, the more likely that programs will fall under the sway of special interest lobbies, Washington deal-making, and excessive generosity.

For all of these reasons, the right answer is to bundle up all of the Federal means-tested programs and send the money back to the states in the form of flexible Block Grants. Under the latter, funds could be used for any form and mix of safety net programs that reflect the varying policy views and political currents among each of the 50 states individually.

Among other things, the resulting Block Grants could be used to systematically pare back to the current bloated level of safety net support funded by Federal taxpayers. We'd propose phasing in cuts in the mean-tested block grant over four years, reaching a maximum of 25% versus current law by 2029. Relative to current baseline spending of $1.15 trillion in FY 2029 that would amount to a $290 billion budget savings.

There is ample reason for a rollback of that magnitude. Thus, in constant FY 2024 dollars of purchasing power, the means-tested programs listed below cost just ***$200 billion*** in FY 1980 when the Reagan Administration attempted to pare back the Welfare State but had risen to ***$985 billion*** by FY 2024.

But this huge ***400%*** increase was not due to an explosion of the poverty population. In fact, the number of persons below the Federal poverty line in 1980 stood at 29.3 million, which figure had risen by only ***25%*** to 36.7 million by 2023. At the same time, the Federal poverty line didn't change in real terms: For a single recipient in 1980 it was *$14,500* in FY 2024 dollars, which is the same as the current level.

Accordingly, constant dollar spending per capita relative to the below-the-poverty population has soared over the last 44 years. In 1980 the figure stood at ***$6,800*** per capita, which was ***47%*** of the $14,500 constant dollar poverty line. By FY 2024, however, that figure was ***$26,850*** per capita or ***185%*** of the poverty line.

Needless to say, if today's far-flung trillion-dollar Little Welfare State was effectively targeted, there wouldn't be anyone under the poverty line in the US at all, let alone the current 37 million. If anything makes a case for defederalizing welfare those figures surely do in spades.

Moreover, the proposed block grant for means-tested welfare at a permanent level of 75% of current law would still be far more generous than the levels that prevailed in 1980. Again, in constant FY 2024 dollars, the proposed block grant in 2029 would total $785 billion, which would amount to ***$21,400*** per capita or about ***150%*** of the current poverty line of $14,500 for a single person.

So with that kind of no-strings attached help from Uncle Sam plus their own tax resources, there is no reason in the world why the states could not fashion a solid safety net. Yet one which is tailored to local conditions, needs, and the flexibility to deviate from today's rigid one-size-fits-all design.

Baseline FY 2029 Spending for Major Means-Tested Welfare Programs:

- Medicaid: $691 billion.
- Obamacare and Child Health Insurance: $120 billion.
- Food Stamps: $113 billion.
- SSI: $71 billion.
- Child Nutrition programs: $40 billion.
- Family Assistance and foster care: $38 billion.
- HUD housing aid and community development programs: $77 billion.
- Total, Means-Tested Welfare: $1,150 billion.
- Savings impact of 25% cut: ***-$290 billion.***
- FY 2029 Block Grant Funding Level: $860 billion.
- 2029 Funding Level In FY 2024 dollars: $785 billion.

Block Grant for Federal Education and Social Services

The scores of narrow categorical grants for elementary and secondary education, job training, and social services will total ***$105.3 billion*** in FY 2025. This immense total provides yet another object lesson in why social spending needs to be thoroughly defederalized and thereby extracted from the clutches of the lobbyists and careerists who prowl the banks of the Potomac.

Back in the day, this sector was a high priority for the Reagan Administration's assault on overspending and Federal waste. In FY 2024 dollars of purchasing power, education, training, and social services spending stood at ***$72 billion*** in the budget left on the table by Jimmy Carter. But by FY 1988 that figure had shrunk to just ***$52 billion*** (FY 2024 $), representing a creditable ***27% reduction.***

When the Gipper left town, however, so did the already tepid GOP fidelity to fiscal austerity. Accordingly, the FY 2025 budget for these programs, which has been shorn of most of the pandemic-era bailouts and free stuff, will clock in at a level ***2X*** the outgoing Reagan budget in constant dollars. In other words, when you have a myriad of narrow programs controlled down in the weeds at the Congressional subcommittee level the tendency for deal-making, log-rolling, and plain old lobbyist campaign graft becomes insuperable. In this case, every vestige of the Reagan spending cuts got washed away with a whole lot of help from GOP appropriators over the decades.

That's why a no-strings Block Grant approach is imperative. The latter can be authorized on a five-year basis with firm annual funding caps while leaving no other reason for annual legislative tinkering and deal-making. Like in the old days of Nixon's General Revenue Sharing program, the spendthrift appropriators of both parties would have nothing to do in the annual funding bills except to approve a large lump sum amount that would then be turned over to thousands of state and local officials on a formula reflecting each state's share of Federal taxes.

In the current instance, we would propose to phase in funding caps that would decline each year until reaching a steady state of 70% of current levels by 2029. Thus, FY 2029 baseline spending of $116.7 billion for this complex of programs would be rolled back to $82 billion, which, in turn, would translate to $76 billion on a constant dollar basis (FY 2024 $).

That's still nearly 50% above the $52 billion outgoing Reagan level and even a tad more than "Big Spender" Jimmy

Carter had in place when he shuffled back to Plains, Georgia in 1981. Besides, blue state liberals would be free to toss in supplements from their own tax dollars if their voters really want more education, training, and social services spending. And this no-strings, formula-based Block Grant would also rid any need for the Federal Education Department, as well as major parts of the Labor Department and HHS, as well.

And, most importantly, a no-strings Block Grant would get Washington politicians and bureaucrats out of the local education business entirely. Under those conditions, we rather suspect that red state officials would give short shrift to today's woke and DEI-based directives and guidelines from Washington bureaucrats, and that even blue state officialdom might find itself called to the carpet by local voters in the event that it persisted with this piddling nonsense.

Stated differently, Block Grants are the antidote to today's unelected but all-powerful administrative state. The latter thrive on control of discretionary money from Uncle Sam–so the solution is to extinguish their power at its fiscal source.

Student Aid and Higher Education

If you don't think student aid and Federal higher education subsidies are utterly out of control, just consider this: In FY 2024 dollars of purchasing power, Federal spending for higher education totaled $26 billion in 1981 but will hit $111 billion by 2029 under the current baseline policy. Yet contrary to this 325% gain, the US college student population has grown only modestly, from 12.5 million in 1981 to 19.2 million at present, and is projected by *Statista* to reach just 20.1 million by 2029.

In this context, the total student population is a fair

denominator because students from all income levels and circumstances get in on the student loan and grant gravy train. For example, nearly 30% of Federal aid goes to students from families with incomes of $100,000 or higher.

Accordingly, Federal aid per college student has soared. In *constant dollars* (FY 2024 $), outlays have risen from ***$2,100*** per student in 1981 to $4,300 per student in the current budget year and to what would be ***$5,500*** per student under baseline spending for FY 2029. There is absolutely no cost-benefit analysis, however, which shows that the American public or economy has gained anything at all from 2.6X more taxpayer largesse per college student.

As it happens, the story of disappearing Reagan cuts is even more evident in the case of student aid. Thus, by 1988 constant dollar spending had fallen from $2,100 to ***$1,450 per student or by 31%***. Again, however, like some kind of fiscal super weed, student aid and higher education funding grew back to $4,300 per student by the current year or *triple* the Reagan 1988 level.

For this reason, we have proposed to consolidate all higher education programs into a no-strings Block Grant, abolish Federally funded student loans entirely, and cut funding by 40% from the FY 2029 baseline. That would generate budget savings of $50 billion, but even then it would be hard to claim undue hardship. After all, even after this huge retrenchment, the constant dollar level of spending would be ***$3,600*** per student in 2029. Back in 1988, a student population of 13.1 million got by on just 40% of this level ($1,450) in real terms per student. Today's 20 million students could do so again, even if it required more self-help from families and student jobs.

Medicare Efficiency Savings

Medicare has become a runaway freight train fiscally. That's mainly owing to the fact that provider payments under the DRG system for hospitals and the MPFS (Medicare Physician Fee Schedule) for doctors are a form of cost-based reimbursement. Accordingly, both cost escalation in the provider marketplace and "coding creep" designed to maximize DRG revenues per patient-stay relentlessly inflate the payment system.

Not surprisingly, the constant dollar (FY 2023 $) cost of Medicare has shot the moon, rising from $131 billion in 1981 to $1.0 trillion in 2023 and is projected at $1.16 trillion under baseline policy by 2029. This ***9X*** growth in real terms over 48 years is not remotely due to the ***2.5X*** increase in the Medicare population from 29 million in 1981 to 65.5 million currently and an estimated 72 million by 2029.

To the contrary, the overwhelming share of the budget cost escalation has been due to relentlessly rising constant dollar cost per Medicare enrollee. Even with the proposed 10% efficiency improvement in Medicare services delivery, projected costs per enrollee by 2029 would be nearly 4X their 1981 level.

Constant Dollar Cost Per Medicare Enrollee (FY 2023 $):

- 1981: ***$4,580.***
- 2023: $15,500.
- 2029 baseline: $18,000.
- 2029 with 10% Efficiency Gain: ***$16,100***

At the end of the day, there is a pretty simple way to achieve a 10% efficiency gain in Medicare payments to providers. To wit, cut reimbursement rates under MPFS for physicians and DRG for hospitals by 10% from current levels. The latter embody the huge inefficiencies that have crept into the system over years and decades of cost-based rate escalation, which could be extracted by the stroke of Uncle Sam's paymaster pen in the short run.

Over time, of course, a new market-based reimbursement system needs to be designed and substituted for the current monument to provider socialism. Starving the beast by an immediate revenue cut, however, might well foster ideas from the providers themselves for a payments mechanism with more of a capitalist flavor, such as converting the Medicare entitlement to a cash basis at the option of the beneficiary.

Indeed, once you get a few million seniors (with not a lot to do otherwise) in the healthcare markets diligently shopping around for medical coverages and services with $16,000 of cash in hand, the amount of system reorganization and efficiency that would materialize would be both far-reaching and effective.

Means Test Social Security and Medicare

As we have indicated the combined cost of Social Security and Medicare comprise the budgetary colossus that envelops 71% of the funding space in the entire entitlements/mandatories budget. To repeat, under the 2029 CBO baseline combined Social Security and Medicare outlays of $3.43 trillion compare to total entitlements/mandatories spending of $4.86 trillion. Yet virtually not a single one of the 70 million plus recipients of these so-called "social insurance" benefits actually earned them during their working lifetime via actuarial "premiums."

That's because Social Security is ultimately a Ponzi Scheme in which all retirees receive substantially more in lifetime benefits than they paid in payroll taxes during their working years.

That's partially due to the progressive wage replacement formulas in the benefits schedule and also due to the fact that lifetime wages are indexed for both inflation and productivity during the period in which workers were paying into the system.

The latter is problematic in its own right because wages earned by a 20-year-old are escalated for both inflation and all productivity gains during the next 47 years of his working lifetime. For instance, if a hypothetical worker had earned ***$5,000*** in 1975 and retired in 2022, those wages under current law would be indexed by **7.01X**, or to ***$35,000*** for purposes of calculating the monthly Social Security benefit.

And if this worker had been a low-wage worker his entire life, the benefit replacement rate for these indexed wages (AIME or average indexed monthly earnings), would be 90%.

Not surprisingly, this structure of wage-indexing and progressive benefit formulas results in a vast disconnect between the *present value* of payroll taxes paid and benefits received. Based on 2021 dollars, here are the benefits-to-taxes ratios for seven typical beneficiary types. As is evident from the table, even high-wage workers receive lifetime benefits equal to 1.5X to 2.0X their lifetime wages on a present-value basis.

Constant Dollar (2021 $) Lifetime Taxes Versus Benefits=Ratio

- Single Low Wage Worker: $195,000 Taxes/$481,000 Benefits=**2.46X.**

- Single Average Wage Worker: $434,000 Taxes/$615,000 Benefits=***1.41X.***
- Single High Wage Worker: $694,000 Taxes/$725,000 Benefits=***1.04X.***
- Married Couple, One Low Wage Worker: $195,000 Taxes/$934,000 Benefits=***4.79X.***
- Married Couple, One Ave. Wage Worker: $434,000 Taxes/$1,162,000 Benefits=***2.68X.***
- Married Couple, One High Wage Worker: $694,000 Taxes/$1,348,000 Benefits=***1.94X.***
- Married Couple, Two Ave. Wage Workers: $867,000 Taxes/$1,303,000 Benefits=***1.50X.***

In short, unless there is endless and substantial growth of the working-age population year after year, which there currently is not, the Social Security/Medicare trust funds will inherently go tilt–and that's especially the case because the system has been pay-as-you-go since the 1960s. That is, there are no real accumulated "pension assets," just intergovernmental IOUs that are not worth the paper they are printed upon.

There is every reason, therefore, for a sweeping reform of the Social Security/Medicare benefit system as part of the DOGE exercise. That's because ultimately the empty trust funds on which these programs now precariously rest amount to a draw on the *general fisc* of the Federal government.

Stated differently, in an overwhelming aspect Social Security and Medicare are *welfare programs,* which we have called the Big Welfare State. They are not actuarial pension benefits in any meaningful sense and therefore must conform to the very first rule of welfare policy. To wit, welfare needs to be rigorously ***means-tested,*** or you end up with levelized socialism of the worst sort.

As it happens, the combined benefits of a top earner at present amount to $70,200 per year, consisting of $15,500 of Medicare value and $54,700 of annual Social Security payments. Presently there are about 14.1 million Social Security/Medicare recipients in the top quintile of beneficiaries, who have total incomes of $75,000 per year or higher.

On a collective basis, therefore, these ***top quintile recipients collect upwards of $1.0 trillion per year of benefits.*** Based on a sliding scale of benefit phase-out geared to total incomes, it is likely that 60% of current costs to these upper-income retirees could be eliminated.

The resulting savings of ***$575 billion*** per year by 2029 would comprise the difference between meeting the DOGE target of $2 trillion of budget savings or falling drastically and unsustainably short. So at the end of the day, there is no alternative to sweeping reform of the Big Welfare State, either.

At the end of the day, of course, the Federal government is not a business. It is mainly a social welfare transfer payment machine in the form of the Welfare State and a global do-gooder and gendarme in the form of the Warfare State. Accordingly, slashing the fat and terminating idle bureaucrats and useless agencies will only get you a small ways–20% even with sweeping cuts–toward the $2 trillion DOGE target.

Stated differently, Washington's runaway spending and borrowing problem is not really a product of inefficiency or fraud, waste, and abuse by any definition of the terms. It is owing to the massive, self-perpetuating Welfare State and Warfare State behemoths that have implanted deep roots on the banks of the Potomac. DOGE's ultimate mission, therefore, is to uproot the policy errors upon which these liberty- and prosperity-killing monsters thrive.

About Brownstone Institute

Brownstone Institute, established May 2021, is a publisher and research institute that places the highest value on the voluntary interaction of individuals and groups while minimizing the use of violence and force, including that which is exercised by public authority.

Published by Brownstone Institute

Austin, Texas

Made in the USA
Las Vegas, NV
13 March 2025

19479359R00095